The 6555th
Missile and Space Launches Through 1970

by Mark C. Cleary

45th Space Wing History Office

Table of Contents

Preface

Chapter I - Foundations of the 6555th: The Post War Legacy
    Section 1 - Post-War legacy Through 1949
    Section 2 - Activities at Holloman, Eglin and Patrick AFB, 1950-1951

Chapter II - MATADOR and the Era of Winged Missiles
    Section 1 - MATADOR Operations Through 1954
    Section 2 - MATADOR and MACE Operations 1955-1963
    Section 3 - LARK, BOMARC and SNARK Operations
    Section 4 - The NAVAHO Program

Chapter III - The 6555th's Role in the Development of Ballistic Missiles
        Section 1 - Ballistic Missile Test Organizations and Commanders
        Section 2 - The Eastern Test Range in the 1950's
        Section 3 - Ballistic Missile Test Objectives
        Section 4 - The THOR Ballistic Missile Program
        Section 5 - The ATLAS Ballistic Missile Program
        Section 6 - The TITAN Ballistic Missile Program
        Section 7 - Organization, Resources and Activities in the 1960's
        Section 8 - The MINUTEMAN Ballistic Missile Development Program

Chapter IV - Taking the High Ground: The 6555th's Role in Space Through 1970
        Section 1 - U. S. Military Space Efforts Through 1960
        Section 2 - ATLAS, THOR and BLUE SCOUT Space Operations
        Section 3 - The TITAN II/GEMINI Program
        Section 4 - The TITAN III Program
        Section 5 - Organizational Changes 1965-1970

Preface

(1st Edition, November 1991)

When I assumed the duties of Chief, ESMC History Office in January 1986, I was completely unaware of the 6555th's contributions to America's missile and space efforts in the 1950s and 1960s. Like most Americans -- indeed, like most people the world over -- I assumed that the National Aeronautics and Space Administration dominated most aspects of the United States space effort after 1958. Within a few months, however, my review of official histories and other government documents in the ESMC archives presented a much different picture of the U.S. space program. I was struck by the appalling "invisibility" of the 6555th. The unit played a pivotal role in the development of missiles and space launch vehicles in the 1950s and 1960s, but, apart from semi-annual historical reports submitted by the 6555th, not one monograph or professional historical study had been written about the 6555th or its efforts at Cape Canaveral. A brief pamphlet entitled "The Story of the 6555th Aerospace Test Wing" was published by the 6555th in 1967, but it hardly did justice to the Wing's many accomplishments up to that time. After I attended to more pressing matters in the office (e.g., the annual history backlog, reorganization of the archives and the establishment of a video documentary library), I studied the unclassified and declassified documents, the histories and the special studies on which this monograph is based. Sadly, many of the semi-annual historical reports had not been looked at since their declassification in the early 1970s.

The present work is a mosaic of the 6555th's history over the first two decades of its existence. It covers the golden age of Air Force missile and space launch vehicle flight tests at Cape Canaveral, and it ends shortly after the 6555th closed out its last ballistic missile flight test program (the MINUTEMAN III) in 1970. (The 6555th was redesignated a Group for the final time in April 1970, which is another reason to consider the year a watershed in the unit's history.) The 6555th's accomplishments since 1970 certainly deserve treatment as well, but their inclusion would have delayed the present monograph's publication by at least a year. Since the future of the 6555th over the next few years is uncertain, we decided in favor of early publication to give the 6555th more time to savor its history before its operations cease.

This monograph could never have been written without the conscientious efforts of Mr. Marven R. Whipple, the Air Force Missile Test Center history staff, and the dozens of officers, airmen and civilians who contributed six-month historical reports on the 6555th's various offices, branches and divisions. Ms. Jan E. Crespino deserves special thanks for the initial formatting and handling the administrative details of publication. I also want to thank Major Richard W. Sirmons for formating the work, assisting with printing, and researching publication alternatives.

The 6555th, Chapter I, Section1

Foundations of the 6555th: The Post-War Legacy

Post War Legacy Through 1949

In the tradition of great military leaders who plan for the future, heed sage advice, and never rest on the laurels of past victories, General Henry H. "Hap" Arnold (Commanding General of U.S. Army Air Forces) began planning the post-war Air Force many months before the end of World War II. As one of airpower's greatest proponents, Arnold knew the value of scientific research -- he felt the Air Force ought to employ "all the scientific minds" it could find and turn their "wondrous" theories into useful tools. The future, he believed, was tied to new technology -- without it, aviation science would stagnate. In 1944, General Arnold asked the famous aerodynamicist, Dr. Theodore von Karman, to develop a prospectus for future Air Force research. Von Karman organized a group of his fellow scientists into the Scientific Advisory Group (later known as the Scientific Advisory Board), and this group produced its initial report, Where We Stand, in August 1945. In light of the Manhattan Project and the then-recent revelations of German missile, rocket engine, jet engine and airframe technologies, Where We Stand's list of concepts for a "revolution" in aerial warfare was not unanticipated. In the interest of clarity, the concepts in the Advisory Group's report can be grouped in the following manner:[1]

1. Airborne task forces will be sustained to strike targets over great distances; aircraft will move at speeds far beyond the velocity of sound.
2. Perfect communication between command organizations and individual aircraft will be established, and operations will proceed regardless of visibility and weather.
3. Unmanned devices will destroy targets several thousand miles distant; small amounts of explosives will cause great destruction.
4. Target-seeking missiles will defend against present-day aircraft; only aircraft or missiles moving at extreme speeds will be able to penetrate enemy territory protected by missile defenses.

The Scientific Advisory Group presented General Arnold with a 33-volume series, Toward New Horizons four months later. This comprehensive survey of research and development options -- with applications to the Air Force of the future -- underscored the Group's belief that the Air Force would have to "draw on the technological potential of the entire nation" to acquire and maintain technological ascendancy over any potential enemy. America's monopoly on atomic weapons could not last indefinitely. (Indeed, the Soviets were only a few years away from acquiring atomic weapons, and both superpowers would have hydrogen [fusion] bombs by 1954.) An aggressive long-range research and development (R&D) program was required to keep abreast of aircraft and missile developments, even if the atom bomb and Strategic Air Command's bombers presented an entirely credible deterrent for many years to come.[2]

 **GENERAL HENRY H. "HAP" ARNOLD**

Shortly before his retirement, General Arnold created an office within the Air Staff for a military deputy for research and development. He also arranged a contract for a staff of civilian scientists and engineers to make a long-range study of intercontinental warfare concepts. The contract was awarded to the Douglas Aircraft Corporation, and it led to the establishment of the Research and Development (RAND) Corporation in 1946. The first Deputy Chief of Staff for Research and Development was Major General Curtis E. LeMay.[3]

Though all three branches of the military expressed an interest in aerodynamic and/or ballistic missile technology after the war, most of their early efforts in this area focused on service-directed experimentation with captured missile hardware or modest research efforts with civilian contractors to develop "home-grown" rocket motors, fuel pumps, guidance systems, etc. This piecemeal effort reflected competition among the military branches, and it defied a succinct statement of long-range research and development objectives. Despite this state of affairs, there were some early hopeful signs: in 1945, the Electronics Division of the Navy's Bureau of Aeronautics was the first agency to suggest the need for a satellite test program. Under Commander Harvey Hall, the Electronics Division discussed the feasibility of such a program with the Guggenheim Aeronautical Laboratories at Cal Tech, and with three civilian contractors -- the Glenn L. Martin Company, North American Aviation, and the Douglas Aircraft Corporation. All four completed their preliminary analyses in early 1946, and their conclusions suggested that a 2,000-pound satellite could be boosted into orbit if the Navy was willing to pay between $5 million and $8 million to develop a sufficiently large rocket to do the job.[4]

Since the services' entire research budget for 1946 was roughly $500 million (i.e., approximately $100 million for the Army Air Forces and $200 each for the Army and Navy), the Navy could not afford to develop a satellite program on its own. On 7 March 1946, Commander Hall and Captain W. P. Cogswell met with three Army Air Force members of the Aeronautical Board to discuss the possibility of a joint Navy-Air Force space program. The Army Air Force reaction was generally positive: though funding might prove a problem, the Army Air Force members agreed to discuss the matter with General LeMay. The satellite proposal was also placed on the Aeronautical Board's agenda for further discussion on 14 May 1946.[5]

To prepare Army Air Force representatives for the joint-service satellite program discussions in May, General LeMay asked for a RAND group study on satellite feasibility on very short notice. A 321-page study was prepared in three weeks and forwarded to the Pentagon two days before the Aeronautical Board's meeting on May 14th. The RAND study concluded that modern technology made a satellite vehicle feasible, though in this instance a 500-pound instrumented payload was proposed instead of a 2,000-pound satellite. Satellites could be applied to military reconnaissance, weather surveillance, communications and missile guidance, but there would be other scientific and commercial benefits as well: the fields of gravitational research, astronomy, bioastronautics and meteorology would profit from the military's space initiatives. The geostationary communications satellite was discussed specifically, and there was even a brief mention of the satellite as a "first step" to interplanetary travel. Engineering aspects were also explored in the study, and RAND's basic calculations concerning vehicle design, fuels, orbital motion, trajectories, developmental requirements and costs were not markedly different from those presented to the American public following the launch of Sputnik more than a decade later.[6]

Unfortunately, shortly after satellites were discussed by the Aeronautical Board's Research and Development Committee on May 14th, the committee reported that no agreement between the Navy and Army Air Force members had been reached. The American scientific community's reaction was not particularly encouraging either, and the Chairman of the Armed Forces' Research and Development Board, Dr. Vannevar Bush, soon denounced ballistic missiles as hopelessly inaccurate and satellites as the vaporings of eminent military men "exhilarated perhaps by a short immersion in matters scientific." The Research and Development Board rejected the satellite proposal on the grounds that it did not support a military requirement, and the Board's Guided Missile Committee refused to fund it. Other missile programs also disappeared in 1947 and 1948, condemned as "too theoretical" or too far removed from existing requirements to warrant funding. Nevertheless, RAND studies on satellites and space vehicles continued, and one week after the U.S. Air Force came into being, Air Force Headquarters asked Air Materiel Command's Engineering Division at Wright-Patterson, Ohio to evaluate the RAND reports for technical and operational feasibility.[7]

While missile studies continued, the expansion of Strategic Air Command demanded much of the Air Force's attention in the late 1940s and early 1950s. In the constant struggle for funding, missile research and development often took a backseat to bomber and tanker force improvements at Air Materiel Command. Despite this fact, missile experimentation was not ignored at Wright-Patterson. On the contrary, when the Pilotless Aircraft Branch was created in 1946, requirements were laid down for many different missiles, as Major General Harry J. Sands, Jr. (USAF, Retired) remembers:[8]

> That was the beginning of the Air Force missile program...getting started wasn't easy, but we knew our missiles could be broken down into four categories: surface-to-surface, surface-to-air, air-to-surface, and air-to-air. It took us a year or so to lay down requirements. We let 26 contracts with 26 contractors just to define our parameters -- we didn't need to spend much money on this...During that first year, all we really wanted the contractors to do was gather knowledgeable people in propulsion, guidance, launching, etc., and see what a reasonable set of missile requirements would be. We reduced the number of primary contractors after that first year, though just about everybody participated through sub-contracts. The (Glenn L.) Martin Company became the primary contractor in short-range, surface-to-surface missiles (e.g., the MATADOR). Boeing got the surface-to-air business with the BOMARC. General Dynamics (i.e., the Consolidated-Vultee Aircraft Corporation) had the ATLAS, and North American had the NAVAHO. Both of those were long-range, surface-to-surface missiles, though the NAVAHO was air-

breathing and the ATLAS was ballistic. Northrop had the SNARK, another long-range, surface-to-surface, air-breathing missile...In the air-to-air business, we wound up with only one contractor: Hughes...In the air-to-surface business, the Bell Company produced the RASCAL...The contractor would be responsible for the requirements, and ultimately responsible for seeing that the pieces came together.

Following the definition of missile requirements, Air Materiel Command started looking for likely places to allow its contractors to launch missiles. Though the Consolidated-Vultee Aircraft Corporation conducted some static rocket engine tests at Point Loma near San Diego, the remnants of this project moved to the White Sands Proving Ground in the spring of 1948. The Air Force also relocated a portion of its JB-2 "buzz bomb" test effort to Holloman Air Force Base at White Sands. The Eglin Air Proving Ground was used for drone aircraft operations, guided bomb experiments and ordnance testing in the late 1940s, but an Air Force detachment also carried out some missile activities at the Navy's Guided Missile Test Center at Point Mugu, California. While Eglin and Wright-Patterson were the organizational "hubs" for many Air Force missile-related projects in the late 1940s, the focus began to shift to other commands and bases in the early 1950s.9

BANANA RIVER NAVAL AIR STATION

One of the first steps in that direction occurred in September 1948, when a deactivated World War II patrol base -- the Banana River Naval Air Station -- was transferred to the Air Force as a base of operations for a joint-service missile range. The Headquarters for the Joint Long Range Proving Ground was set up there on 10 June 1949, but joint management of the Proving Ground proved unwieldy, and the Headquarters was replaced by the Air Force's Long Range Proving Ground Division on 16 May 1950. Under the Air Force's management, the Proving Ground started building launch complexes, missile processing facilities and instrumentation sites at Cape Canaveral and elsewhere on the Florida mainland. Within a decade, the Proving Ground evolved into the Atlantic Missile Range (later known as the Eastern Test Range), and its instrumentation sites extended from Cape Canaveral to range ships and island stations reaching all the way to Ascension Island in the South Atlantic. While ordnance activities remained at Eglin, missile units gravitated toward Cape Canaveral, the future home of the 6555th Guided Missile Wing .10

B-17 DRONE TAKE-OFF
Control Aircraft At Top
Right

The 6555th Guided Missile Wing's predecessors included the 1st Experimental Guided Missiles Group, the 550th Guided Missiles Wing and the 4800th Guided Missile Wing. The 1st Experimental Guided Missiles Group was activated at Eglin Field, Florida on 6 February 1946. Pursuant to an order from the

War Department (dated 25 January 1946), the Commanding General of the Army Air Forces Center at Eglin Field was directed to activate the Headquarters, 1st Experimental Guided Missiles Group, the 1st Experimental Guided Missiles Squadron and the 1st Experimental Air Service Squadron. The total authorized strength for the three organizations was 130 officers, one warrant officer and 714 enlisted men. Eglin's commander was directed to supply manpower for the units from his own resources, but, given the recent postwar demobilization, his ability to do so was extremely limited. Of the four officers and three airmen assigned to the Group on February 6th, the Group Commander and two lieutenant colonels were on temporary duty with an air instrumentation and test requirement unit supporting Project CROSSROADS in the Pacific. Colonel Harvey T. Alness did not actually assume command of the Group until early August, shortly after he and lieutenant colonels William B. Keyes and Richard A. Campbell brought two B-17 drones to Muroc Air Force Base, California from Hawaii following the CROSSROADS operation in July 1946. In the interim, Captain Wheeler B. Bowen (the only officer on station when the 1st was activated) remained in temporary command until Major Frederick M. Armstrong Jr. joined the Group on February 15th. Lieutenant Colonel Wesley Werner replaced Armstrong as Commander on May 13th, and he continued in command until Colonel Alness returned in August 1946.11

GROUND CONTROL UNIT FOR B-17 DRONE

During its first year of operation, the 1st Experimental Guided Missiles Group operated out of Eglin's Auxiliary Field #3. Personnel attended technical schools or supported other Air Proving Ground units, but apart from receiving nationwide attention in January 1947 for completing a drone flight from Eglin to Washington D.C. on a simulated bombing mission, the Group received little notice in its own right. Without higher supply and personnel priorities, very little else could be accomplished. The situation began to change in March 1947, when the Group moved to Eglin's main base and received its first series of test projects. The Group was given the JB-2 -- an American version of the German V-1 flying bomb -- and it got involved in VB-6 FELIX, VB-3 RAZON, and VB-13 TARZON guided bomb activities.

Though most of the Group's efforts were devoted to "on-the-job" training and providing assistance to contractors who launched those weapons, the 1st began implementing its mission, which included: 1) developing tactics and techniques for guided missile operations, 2) training personnel and testing equipment used in guided missile organizations, 3) developing requirements and standards for the employment of guided missiles, and 4) conducting functional and tactical tests of new guided missiles to determine their operational suitability (i.e., readiness for adoption by the armed forces). The Group also began providing observers for guided missile tests at laboratories and factories, including those programs sponsored by the Army and Navy.12

 *RAZON BOMB BEING LOADED*

1952

Though preparations for another atomic test (Project SANDSTONE) engaged most of the Group's resources from July 1947 through June 1948, the 1st Experimental Air Service Squadron picked up responsibility for drone aircraft bombing tests (e.g., Project BANSHEE) and conducted a limited number of JB-2 and VB-6 tests during that period. The 1st regrouped its activities after Project SANDSTONE, and it spent several months preparing a detachment to depart for cold weather testing of the JB-2 in Alaska in November 1948. RAZON and TARZON bomb tests were underway by the end of the year.13

During the last seven months of its existence, the 1st Experimental Guided Missiles Group either supervised or participated in eleven different missile-related projects. In addition to the on-going JB-2, RAZON, TARZON, FELIX and BANSHEE projects, the Group had a detachment in training at Point Mugu to handle and operate the Navy's LARK surface-to-air missile. The Group also provided a detachment to support the MATADOR project at Holloman Air Force Base, New Mexico. The Group's other projects included preparation for the GREENHOUSE atomic test (conducted in 1951), drone aircraft "ditching" tests (to test structural weaknesses) and drone aircraft support for high-altitude incendiary ammunition tests and infrared radiation experiments.14

On 20 July 1949, the 1st Experimental Guided Missiles Group was deactivated, and it was replaced by the 550th Guided Missiles Wing on the same date. At the time of its deactivation, the 1st had 97 officers and 523 airmen assigned to it. Those people were transferred to the 550th, and, in general, they were assigned duties identical to their tasks in the old Experimental Guided Missiles Group. However, unlike the 1st, the 550th was a wing, and it had four squadrons to carry out its functions:15

1. The 550th Headquarters and Headquarters Squadron (drawn from the 1st Group's Headquarters).
2. The 1st Guided Missiles Squadron (composed mainly from personnel taken from the deactivated 1st Experimental Guided Missile Squadron).
3. The 2nd Guided Missiles Squadron (manned by personnel from the 1st Experimental Guided Missile Squadron and the 1st Experimental Air Service Squadron).
4. The 550th Maintenance Squadron, Guided Missiles (composed of the remaining portion of the 1st Experimental Air Service Squadron, which was merely redesignated on 20 July 1949).

The 6555th, Chapter I, Section 2

Foundations of the 6555th: The Post-War Legacy

Activities at Holloman, Eglin and Patrick AFB
1950-1951

The Wing's mission contained the essential elements of the old 1st Group's mission, but emphasis was placed on supervision and evaluation of guided missile service tests as opposed to pure experimentation. (Consequently, the word "experimental" was omitted from the Wing's name and its squadrons' designations). Like its predecessor, the 550th Guided Missiles Wing had detachments in tenant status at Holloman Air Force Base and the Navy's Guided Missile Test Center at Point Mugu. By December 1949, the detachment at Holloman was authorized 25 officers and 45 airmen, and the Point Mugu detachment had 11 officers and 30 airmen in place. While the Holloman detachment continued to assist the Glenn L. Martin Company with developmental testing of the MATADOR (i.e., it witnessed test firings and reported on the results), the Point Mugu detachment completed its LARK training and moved to the Joint Long Range Proving Ground in early January 1950.[16]

At Eglin, the 1st Guided Missiles Squadron was assigned air-to-surface missiles and guided bombs (e.g., TARZON) and the 2nd Guided Missiles Squadron worked with surface-to-surface missiles and aircraft drones. During the first ten months of its existence, the 550th Guided Missiles Wing also continued its predecessor's earlier preparations to support Project GREENHOUSE with drone aircraft, but additional drones and personnel were assigned to other Air Proving Ground units during this period as well. By January 1950, the Air Proving Ground decided this piecemeal operation ought to be consolidated, and it recommended the establishment of a separate and permanent drone squadron. Personnel from the 2nd Guided Missiles Squadron were subsequently transferred to a new unit -- the 3200th Drone Squadron, 3200th Proof Test Group -- in May 1950. While the 3200th Drone Squadron remained under the 550th for administrative purposes, its operations were essentially divorced from the 550th's missile activities when the 3200th moved to Auxiliary Field #3. The 2nd Guided Missiles Squadron was placed on inactive status after the transfer, but it was revived at Holloman Air Force Base on 25 October 1950 when the 550th's detachment out there was discontinued. As the 2nd Guided Missiles Squadron Commander at Holloman, Captain John A. Evans inherited the old detachment's manpower and gained 40 airmen from other Wing resources. This brought the Squadron's strength to 17 officers and 114 airmen (out of the 550th's total complement of 201 officers and 816 airmen).[17]

While missile testing continued in 1950 and 1951, the Air Force reorganized the oversight of its research and development program under the auspices of a new major agency -- the Air Research and Development Command (ARDC). The new command was activated on 23 January 1950 with Major

General David M. Schlatter as its commander. By April 1951, Wright-Patterson's research and development agencies, various laboratories, Edwards Air Force Base and Holloman Air Force Base had been transferred from Air Materiel Command to ARDC. By the end of 1951, ARDC's principal field components included the Wright Air Development Center at Wright-Patterson Air Force Base, the Air Force Flight Test Center at Edwards Air Force Base, the Air Force Armament Test Center at Eglin, and the Air Force Missile Test Center at Patrick Air Force Base (about 15 miles south of Cape Canaveral). The Holloman Air Development Center was established at Holloman Air Force Base in 1952.[18]

MAJOR GENERAL DAVID M. SCHLATTER

Though research and development became focused under ARDC, Air Materiel Command still had an important role to play in the acquisition of new Air Force weapon systems. Obviously, ARDC could develop a weapon system to the point where it was deemed suitable for operations, but it was Air Materiel Command's job to bring the new system into the Air Force inventory and address all the production problems that typically entailed.* The Weapon System Project Office (WSPO) served as a "bridge" between ARDC's activities and Air Materiel Command's procurement effort, and it administered and controlled individual weapon system programs. Air Materiel Command continued to direct initial procurement of weapon systems until Air Force Systems Command assumed that responsibility in April 1961.[19]

Given the course of the reorganization, the reassignment of missile units from the Air Proving Ground to the Long Range Proving Ground Division in 1950 was quite understandable. The 1st Guided Missiles Squadron and other missile units did not wind up at Cape Canaveral just because they needed a longer range to test their missiles. If that had been the only concern, missile units could have continued as tenants at Patrick and merely reported to a higher headquarters at Eglin. The longer test range was an important consideration, but missile units were assigned to the Long Range Proving Ground Division because it was a new intermediate headquarters specifically designed to support guided missile test programs that were emerging as weapon systems in their own right. As ARDC refocused the Air Force's R&D effort, it made Cape Canaveral the principal launch site for surface-to-surface and surface-to-air missiles. For the most part, "armaments" (including air-to-air and air-to-surface missiles) remained at Eglin or Holloman.

As ARDC took over responsibility for missile research and development, the Air Force directed the 550th Guided Missiles Wing in late November 1950 to move its headquarters, the 550th Maintenance Squadron and the 1st Guided Missiles Squadron to the Long Range Proving Ground. On December 6th, the 550th activated a detachment at Patrick Air Force Base and assigned it to the 3rd Guided Missiles Squadron, Interceptor to coordinate the movement to Patrick. The 550th also activated a detachment at Eglin (consisting of 9 officers and 25 airmen from the 1st Guided Missiles Squadron) to run suitability tests on the unfinished TARZON project. The 1st Guided Missiles Squadron Commander, Major Henry B. Sayler, remained at Eglin to command the TARZON detachment, but the rest of his squadron was transferred to the 3rd Guided Missiles Squadron, Interceptor, raising that unit's manpower to 18 officers and 153 airmen. The move was accomplished between 12 and 18 December 1950. Its work completed, the Patrick detachment was discontinued. In all, the 550th moved approximately 30 officers and 170 airmen to Patrick in December. The 550th's new commander, Colonel George M. McNeese, supervised the Wing's departure, but Lieutenant Colonel Jack S. DeWitt completed the operation from 14 through 18 December 1950.[20]

COLONEL GEORGE M. MCNEESE

COLONEL JACK S. DEWITT

LT. COLONEL HENRY B. SAYLER

The Long Range Proving Ground Division inactivated the 550th and its squadrons on 29 December 1950 and it replaced those units with the 4800th Guided Missile Wing, the 4802nd Guided Missile Squadron, and the 4803rd Guided Missile Squadron on 30 December 1950. Colonel McNeese assumed command of the new Wing, and he appointed Lieutenant Colonel Jack S. Dewitt as his Deputy Commander, Major Theodore H. Runyon as his Deputy for Operations, and Major Robert Maloney, Jr. as the Deputy for Materiel. Major Hamilton commanded the 4803rd until he was reassigned to the Pentagon in March 1951, at which time Lieutenant Colonel Henry B. Sayler assumed command of the Squadron. Orders were also cut in March to move the 4802nd Guided Missile Squadron (and Project MATADOR) from Holloman to Patrick. This movement was completed when the 4802nd's commander, Lieutenant Colonel John C. Reardon, reported in with his squadron on 12 April 1951. Having inherited all the missile-related portions of the 550th's mission, the 4800th had the following resources to carry out its responsibilities:[21]

- Headquarters & Headquarters Squadron: 20 officers and 46 airmen
- 4802nd Guided Missile Squadron: 19 officers and 114 airmen
- 4803rd Guided Missile Squadron: 19 officers and 164 airmen
- Detachment 1 (at Eglin): 8 officers and 29 airmen

AERIAL VIEW OF CAPE CANAVERAL - 1955

AERIAL VIEW OF PATRICK AFB - 1955

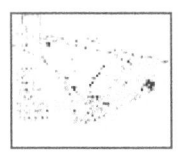
MAP OF CAPE CANAVERAL

As mentioned earlier, the 550th's LARK detachment was transferred from Point Mugu to the Long Range Proving Ground in January 1950, so a discussion of the 4803rd's launch operations at Cape Canaveral must include some mention of the LARK detachment and the 3rd Guided Missiles Squadron's activities. Initially, the detachment thought it would be able to launch its first LARK at the Cape in March 1950, but the lack of Range facilities -- even for the LARK -- convinced detachment authorities that the first LARK launch at Cape Canaveral would have to wait until after BUMPER 8 and BUMPER 7 (i.e., the last two test launches in the Army's BUMPER project) were completed in the summer of 1950. On 1 September 1950, the 3rd Guided Missiles Squadron inaugurated an official training course for guided missile technicians, and the 3rd's operations officer conferred with Range authorities on the same date concerning launch procedures, Range interference control, instrumentation support, and a tentative schedule for the first series of LARK operations. Hurricane Able delayed the first LARK launch in mid-October, but three LARKs were launched successfully by the 3rd Guided Missiles Squadron on October 25th, October 26th and November 22nd 1950. Six more LARKs were launched during the first six months of 1951 as part of the training program for the 4803rd Guided Missile Squadron and its successor, the 6556th Guided Missile Squadron.[22]

LARK MISSILE ON ZERO-LENGTH LAUNCHER - 1950

FINAL INSPECTION OF LARK PRIOR TO ELEVATION FOR LAUNCH - 1951

 LAUNCH OF BUMPER 8 - Cape Canaveral, 24 July 1950

 LAUNCH OF BUMPER 7 - Cape Canaveral 29 July 1950

In similar fashion, the 4802nd Guide Missile Squadron's operations were a continuation of the Holloman Detachment's activities and the 2nd Guided Missiles Squadron's MATADOR training before the 4802nd moved to Patrick. The 550th's Holloman Detachment had been activated on 7 November 1949 to train Air Force personnel on the MATADOR while the Glenn L. Martin Company conducted experiments on the missile and its zero-length launcher.* Military involvement in the MATADOR project amounted to little more than "on-the-job training," but Martin launched 22 MATADOR test vehicles (including 15 dummy missiles) at Holloman before the company moved the MATADOR to Cape Canaveral in the spring of 1951. While the 4802nd "assisted" on some of those launches, the move to Patrick occupied much of its time in early April, and the Squadron was on station only a few weeks before it was redesignated the 6555th Guided Missile Squadron in early May. Nevertheless, thanks to the 4802nd's efforts, the 6555th was prepared to present an 81-hour orientation course on the MATADOR within days of the Squadron's redesignation. As the Glenn L. Martin Company prepared to launch its first MATADOR from Cape Canaveral on 20 June 1951, the 6555th Guided Missile Squadron trained to assemble, check out, launch, and control MATADORS scheduled for later service testing and operational experimentation. The 6555th was also tasked with supervising instruction for Tactical Air Command's first two MATADOR squadrons. Those squadrons were activated on 1 October 1951 and 10 January 1952, and they were assigned to the 6555th Guided Missile Wing subsequently.[23]

 MATADOR ASSEMBLY AREA - May 1951

 PREPARATIONS FOR MATADOR LAUNCH - 1951

 PREPARATIONS FOR MATADOR MOTOR TEST - 1951

After little more than four months of operation, the 4800th Guided Missile Wing and its squadrons were redesignated as part of the Long Range Proving Ground Division's assignment to the Air Research and Development Command. Until 14 May 1951, the Division had been a separate operating agency under Air Force Headquarters. Following its assignment to ARDC, the Division was accorded "numbered air force" status, and almost all of its subordinate units were given 65XX-series designations. In keeping with ARDC's policy of designating its intermediate headquarters as "centers," the Long Range Proving Ground Division became the Air Force Missile Test Center (AFMTC) on 30 June 1951. The Air Research and Development Command assigned Holloman Air Force Base to AFMTC on 3 July 1951, but this action had no immediate effect on guided missile operations at Patrick or Cape Canaveral. Detachment 1 continued to support RAZON and/or TARZON operations at Eglin through the end of July 1951. Its task completed, Detachment 1 was discontinued on 9 August 1951, and its personnel were sent back to the 6555th Guided Missile Wing at Patrick.[24]

These, then, were the events preceding and overlapping the creation of the 6555th Guided Missile Wing: with SNARK, NAVAHO, BOMARC and other major missile projects on the planning horizon, the 6555th prepared to expend its activities in the 1950s and 1960s to develop a military or "blue suit" launch capability for a whole host of tactical and strategic missile weapon systems. In many instances, operational suitability testing had to go hand-in-hand with missile training to insure that the Air Force did not buy a weapon that was too complicated, fragile or unreliable to operate or maintain in the field. Launch vehicle modifications, ground and flight support equipment, assembly and checkout procedures, safety standards and instrumentation requirements had to be thoroughly understood, checked, and verified with range authorities before every launch. No doubt the challenges appeared daunting, but the 6555th was not operating in a vacuum -- it capitalized on its predecessors' work and relationships with contractors at Eglin, Holloman and Patrick. Though new, the 6555th was sustained by the Air Force Missile Test Center, an organization created to support missile projects as future weapon systems in their own right. Under the Air Research and Development Command's direction, some missile projects were destined to evolve at Cape Canaveral into long-lived programs with higher visibility and better funding than they had enjoyed under Air Materiel Command. Thus, the 6555th began to take the measure of General Arnold's future Air Force, at least as far as missiles and space vehicles were concerned. Though the 6555th was hardly alone in this effort, it had a important role to play in the entire exercise.

The 6555th

Chapter One Footnotes

Aeronautical Board
The Aeronautical Board was jointly staffed by the Army Air Forces and the Navy Bureau of Aeronautics. The three officers at the March 7th meeting were Major General Hugh J. Knerr, and Major General H. W. McClellan and Brigadier General William L. Richardson. General Richardson eventually became the first Air Force commander of the Long Range Proving Ground Division (LRPGD) and its successor, the Air Force Missile Test Center (AFMTC).

Other missile programs also disappeared
In December 1946, the guided missile budget for fiscal year (FY) 1947 was reduced from $29 million to approximately $13 million. Eleven missile projects were eliminated, and five more were terminated in May 1947. By the summer of 1947, only the left-overs of the Air Force's Consolidated-Vultee long-range ballistic missile project and eight other missile programs remained. They included two identifiable ballistic missile efforts (e.g., the Navy's Viking project and the Army's Redstone), but, apart from rocket motor research, the Air Force's missile projects centered on airborne tactical missiles and air-breathing winged missiles like the MATADOR.

White Sands Proving Ground
White Sands was a 125-mile-long range set up in 1945 in a high valley north of El Paso, just across the Texas-New Mexico state line. Though the range was only 41 miles wide on the average, it was adequate for WAC-Corporal and V-2 launches. Following the arrival of V-2 components in the summer of 1945, the Army (with the indispensable support of German rocket scientists who had worked on the V-2 at Peenemunde) began launching V-2s from White Sands in early 1946.

Project CROSSROADS
This effort involved directing remote-controlled B-17 drone aircraft into radioactive areas to collect air samples shortly after an atomic test.

VB-6 FELIX, VB-3 RAZON, and VB-13 TARZON
The FELIX was an air-to-surface guided bomb equipped with a heat-seeking guidance system. The RAZON and TARZON were 1,000-pound and 12,000-pound high-explosive bombs whose tail assemblies were modified to allow a bombardier to radio-control their trajectories (within certain limits) following the bombs' release from an aircraft.

1st Experimental Guided Missiles Group
550th Guided Missiles Wing

Colonel John R. Kilgore, who had been in command of the 1st Group since 13 August 1947, relinquished his command upon his unit's deactivation. Colonel Thomas J. Gent, Jr. assumed command of the 550th Guided Missiles Wing on the date the unit was activated.

detachments
The detachment at Point Mugu was formally redesignated the Headquarters, 550th Guided Missiles Wing Detachment on 21 July 1949, but the Holloman detachment was not formally redesignated until 15 November 1949. This was apparently a clerical oversight, since the Holloman detachment had been in place at Alamogordo, New Mexico before the 550th Guided Missiles Wing was activated.

Air Research and Development Command
On behalf of the Air Force Chief of Staff, General Hoyt S. Vandenberg, Dr. von Karman had asked Dr. Louis N. Ridenour in 1949 to chair a committee to study Air Force research and development activities. The Ridenour Committee submitted its report in September 1949, and this report recommended the creation of a research and development command in addition to a position on the Air Staff for a Deputy Chief of Staff for Research and Development. Major General Orval A. Anderson also directed an Air University study on the subject in 1949, and it echoed the Ridenour report, but in stronger terms: research and development ought to be removed from Air Materiel Command and vested in a single agency for research and development.

weapon systems
The term "weapon system" became part of Wright-Patterson's vocabulary at least several years before the creation of ARDC. Major General Harry J. Sands, Jr. recalled using the "systems approach" for missile development and procurement in the Pilotless Aircraft Branch in the late 1940s. A weapon system was formally described as "an instrument of combat...together with all related equipment both airborne and ground based, the skills necessary to operate the equipment, and the supporting facilities and services required to enable the instrument of combat to be a single unit of striking power in its operational environment." The systems approach considered all the elements of a weapon system when requirements were set down on paper.

Air Materiel Command's job
As Deputy Chief of Staff for Materiel, Lieutenant General Orval R. Cook was given responsibility (within the Air Staff) for overall supervision of Air Force R&D in September 1953. In an effort to improve weapon systems management in ARDC and Air Materiel Command, General Cook formed an advisory group to investigate the concept of "cradle to grave" procurement (i.e., detailed planning for research, development, testing, producing, maintaining, repairing and -- ultimately -- disposing of a weapon system). A key feature of this concept was the "fly before you buy" approach, which insured that an initial production run of aircraft or missiles would be thoroughly tested and declared operationally suitable before the Air Force committed itself to full-scale production and deployment of a weapon system.

3rd Guided Missiles Squadron, Interceptor
The 3rd Guided Missiles Squadron, Interceptor had been activated on 1 July 1950, apparently replacing the 550th's missile detachment at the Long Range Proving Ground. A deactivation order (dated 1 August 1950) indicates that the Detachment, Headquarters and Headquarters Squadron of the 550th Guided Missiles Wing at the Long Range Proving Ground was not discontinued until 1 August, but histories of the 1st and 3rd Guided Missiles Squadrons and the 550th Guided Missiles Wing indicate that Major Joseph H. Hamilton (the detachment commander) assumed command of the 3rd Guided Missiles Squadron, Interceptor on either July 1st or July 6th. In any event, the detachment's people, records and equipment were transferred to the 3rd Guided Missiles Squadron. The Squadron's initial muster was 13 officers and 44 airmen and one other officer and 11 airmen were attached to the Squadron to set up a guided missiles school for Air Training Command.

Colonel George M. McNeese
Colonel McNeese frequently assumed command temporarily during Colonel Thomas J. Gent's trips to the 550th's units at Holloman and Patrick in the summer of 1950. McNeese finally assumed command in his own right on 23 October 1950.

inactivated the 550th
As the squadrons' numbers suggest, the old 2nd Guided Missiles Squadron stationed at Holloman became the 4802nd Guided Missile Squadron, and the 3rd Guided Missiles Squadron, Interceptor became the 4803rd Guided Missile Squadron. The 550th Maintenance Squadron was also inactivated on December 29th, but the 550th's movement order to Patrick listed only one officer and one enlisted man from the 550th Maintenance Squadron. It is safe to assume that the rest of the 550th Maintenance Squadron's personnel had been transferred to the 3200th Drone Squadron or some other unit at Eglin before the 550th Guided Missiles Wing departed for Patrick in December.

Lieutenant Colonel Henry B. Sayler
Sayler's detachment at Eglin was not mentioned in the Long Range Proving Ground Division's inactivation order, but it was "established" by the 4800th as Detachment "A" Headquarters & Headquarters Squadron, 4800th Guided Missiles Wing on 30 December 1950. The Wing amended the order on January 26th and made the detachment "Detachment 1".

LARK
The LARK was developed by the Fairchild Aircraft Company during World War II as a Navy anti-aircraft missile. With a range of 35 miles and a speed of 300 knots per hour, the 173-inch long LARK was adopted by the Air Force as a training vehicle for personnel who would later become involved with Project BOMARC at Cape Canaveral. The first LARKs fired at Point Mugu required a 450-foot-long ramp, but a zero-length launcher was used with the LARKs fired at Cape Canaveral. Range support requirements were very modest, even by early 1950s standards.

BUMPER 8 and BUMPER 7
Toward the end of 1946, the Army Ordnance Corps became interested in the concept of a "step-

rocket." It asked the General Electric Company to mount a WAC-Corporal missile atop of a German V-2 rocket and launch a series of those hybrid "Bumper" vehicles at the White Sands Proving Ground. Six BUMPER missiles were launched at White Sands in 1948 and 1949, and those flights verified the satisfactory operation of both missile stages and their separation system. Two more flights were planned with relatively low, flat trajectories (i.e., less than 150,000 feet in altitude), but White Sands was too short to accommodate them. The Long Range Proving Ground had the requisite length (250 miles), so BUMPERs 8 and 7 were launched from Cape Canaveral on 24 July and 29 July 1950 respectively. The General Electric Company was responsible for launching the vehicles, and the Army's Ballistic Research Laboratories (Aberdeen Proving Ground, Maryland) provided instrumentation support. Among the Army and Air Force units that supported the BUMPER flights from the Cape, the 550th Guided Missiles Wing provided several aircraft and crews to monitor the Range for clearance purposes. The Long Range Proving Ground Division provided overall coordination and range clearance.

training course
The course covered the LARK's propulsion and guidance systems. The first graduating class consisted of a dozen Air Training Command personnel who returned to their parent command to establish a school for guided missile technicians.

MATADOR
The MATADOR B-61A "pilotless bomber" was just emerging from its developmental stage in 1951. It was designed as a 650-mile-per-hour winged tactical missile built to carry a 3000-pound conventional or nuclear warhead a distance of approximately 500 miles. The MATADOR utilized a solid propellant rocket bottle as a Rocket Assisted Takeoff (RATO) system to lift itself into the air from a "roadable" zero-length launcher. After the rocket burned out and dropped off, the MATADOR was powered to its target by an Allison J-33 turbojet engine. Tests in the early 1950s included the development of two different guidance systems: the MATADOR Automatic Radar Command "MARC" system and the Short Range Navigation Vehicle "SHANICLE" microwave system. Both systems required ground stations to control the missile's airborne guidance hardware. While the early test version of the missile measured 34 feet, 7 inches long and had a wing span of 23 feet, 4 inches, the production models were 39.6 feet long and measured 28.7 feet from wing-tip to wing-tip.

The 6555th

Chapter One Endnotes

1. through 3. Stanley, Dennis J. and John J. Weaver, An Air Force Command for R&D, 1949-1976: The History of ARDC/AFSC, AFSC History Office, undated, pp. 2, 4, 6-8.

4. Perry, Robert L., Origins of the USAF Space Program 1945-1956, Space Systems Division History Office, 1961, pp. 9,10; Interview, Major General Harry J. Sands, Jr., USAF Ret., with Mark C. Cleary, 30 April and 2 May 1990, pp. 10-13.

5 and 6. Perry, Origins, pp. 10, 12, 15.

7. Ibid., pp. 19-21.

8. Interview, Sands, pp. A2-25 through A2-27; Stanley, An Air Force Command for R&D, p. 8.

9. Ley, Willy, Rockets, Missiles and Men in Space, N.Y. Viking Press, 1968 Edition, pp. 225, 232, 325, 326; Ferris, Robert G. and Russell D. Roth, The Air Proving Ground's Role in the Air Force Missile Program: Pioneering and Supporting the Development of the Weapons of Tomorrow, Air Proving Ground Center Historical Division, undated, pp. 8-12; Historical Resume of the 4800th Guided Missile Wing, February 1946 - April 1951, pp. 3, 4, 9, 12, 13; Interview, Sands, pp. 25, 27, A2-30.

10. Ltr, HQ USAF, "Redesignation and Change in Assign(ment) of the AFD, JLRPG," 5 May 1950; LRPGD History, 1 January - 30 June 1950, pp. 1-5; ESMC History, 1 October 1988 - 30 September 1989, Vol. I, p. 2; AFMTC History, 1 July - 31 December 1957, Vol. I, p. 10; Historical Resume of the 4800th Guided Missile Wing, February 1946 - April 1951, pp. 10-14.

11. Historical Resume of the 4800th Guided Missile Wing, February 1946 - April 1951, pp. 1-3; Ferris, The Air Proving Ground's Role, p. 14; History of the 550th Guided Missiles Wing, 1 October - 31 December 1950, p. 3.

12. Historical Resume of the 4800th Guided Missile Wing, February 1946 - April 1951, pp. 2, 3; Ferris, The Air Proving Ground's Role, pp. 8 and 9; Pamphlet, 6555th Aerospace Test Wing, "The Story of the 6555th Aerospace Test Wing," o/a July 1966, p. 3.

13. Ferris, The Air Proving Ground's Role, p. 9.

14. Historical Resume of the 4800th Guided Missile Wing, February 1946 - April 1951, pp. 4-6.

15. General Order Number 24, HQ Air Proving Ground, "Activation of the 550th Guided Missiles Wing," 19 July 1949; Historical Resume of the 4800th Guided Missile Wing, February 1946 - April 1951, pp. 3, 7, 8; Ferris, The Air Proving Ground's Role, p. 10; General Order Number 1, HQ 550th Guided Missiles Wing, "Assumption of Command,"20 July 1949.

16. General Order Number 2, HQ 550th Guided Missiles Wing, 21 Jul 1949; General Order Number 35, HQ Air Proving Ground, 30 November 1949; Historical Resume of the 4800th Guided Missile Wing, February 1946 - April 1951, pp. 7, 9, 10; Movement Orders, HQ Air Proving Ground, "Movement Orders, Detachment Hq. and Hq. Sq., 550th Guided Missiles Wing," 9 December 1949; General Order Number 3, HQ 550th Guided Missiles Wing, "Activation of 550th Guided Missiles Wing Detachment," 7 November 1949; LRPGD History, 1 July - 31 December 1950, p. 9.

17. Ferris, The Air Proving Ground's Role, pp. 10, 13; Historical Resume of the 4800th Guided Missile Wing, February 1946 - April 1951, pp. 11, 12; History of the 550th Guided Missiles Wing, 1 October - 31 December 1950, pp. 4, 32; General Order Number 8, HQ 550th Guided Missiles Wing, 19 October 1950; History of the 1st Guided Missiles Squadron ASM and the 3rd Guided Missiles Squadron, Interceptor, 1 October - 31 December 1950, pp. 4, 8.

18. Stanley, An Air Force Command for R&D, pp. 10-12, 17.

19. Ibid., pp 19-21, 44.

20. General Order Number 13, HQ 550th Guided Missiles Wing, "Activation of 550th Guided Missiles Wing Detachment," 20 Nov 1950; Historical Resume of the 4800th Guided Missile Wing, February 1946 - April 1951, pp. 13, 14, 15; Ferris, The Air Proving Ground's Role, p. 15; General Order Number 3, HQ 550th Guided Missiles Wing, "Discontinuance of Detachment, Hq. and Hq. Squadron, 550th Guided Missiles Wing, Long Range Proving Ground Air Force Base, Florida," 1 August 1950; History of the 1st Guided Missiles Squadron ASM and the 3rd Guided Missiles Squadron, Interceptor, 1 October - 31 December 1950, pp. 5, 6, 16; History of the 550th Guided Missiles Wing, pp. 27, 28, 29, 33; DAF Movement Directive, "Movement Directive, 550th Guided Missiles Wing," 27 November 1950; LRPGD History, 1 July - 31 December 1950, p. 13; General Order Number 14, HQ 550th Guided Missiles Wing, "Activation of Detachment Headquarters & Headquarters Squadron 550th Guided Missiles Wing," 6 December 1950; Special Order Number 240, HQ 550th Guided Missiles Wing, 6 December 1950; General Order Number 10, HQ 550th Guided Missiles Wing, "Assumption of Command," 23 October 1950.

21. General Order Number 56, HQ LRPGD, "Inactivation of the 550th Guided Missiles Wing and Establishment of the 4800th Guided Missile Wing," 21 December 1950; General Order Number 2, HQ 4800th Guided Missile Wing, 30 December 1950; Historical Resume of the 4800th Guided

Missile Wing, February 1946 - April 1951, pp. 14, 15, 16; DAF Movement Directive, "Movement Directive, 550th Guided Missiles Wing," 27 November 1950; History of the 4800th Guided Missile Wing, January 1951, p. 1; General Order Number 3, HQ 4800th Guided Missile Wing, 26 January 1951; Movement Order Number 1- 51, 4800th Guided Missile Wing, 20 March 1951.

22. LRPGD History, 1 January - 30 June 1950, pp. 72, 130; LRPGD History, 1 July - 31 December 1951, pp. 39, 41, 142-144, 148, 154-157; Historical Resume of the 4800th Guided Missile Wing, February 1946 - April 1951, pp. 9. 13; Ferris, The Air Proving Ground's Role, pp. 12, 13.

23. Historical Resume of the 4800th Guided Missile Wing, February 1946 - April 1951, pp. 9, 13; TAC Fact Sheet, "LARK," undated; Patrick Office of Information Services Fact Sheet, undated, 3 March 1955; LRPGD History, 1 January - 30 June 1951, pp. 72, 119; LRPGD History, 1 July - 31 December 1950, p. 158; 6555th Guided Missile Wing History, November - December 1951, pp. 3, 36; AFMTC History, 1 January - 30 June 1953, pp. 146-148; General Order Number 24, HQ LRPGD, 14 May 1951; AFMTC History, 1 January - 30 June 1952, p. 428.

24. LRPGD History, 1 January - 30 June 1951, pp. 22-25; 6555th Guided Missile Wing History, May 1951, pp. 5-7; 6555th Guided Missile Wing History, July 1951, pp. 4-7; General Order Number 8, HQ ARDC, 14 May 51; General Order Number 24, HQ LRPGD, 14 May 1951; General Order Number 3, HQ AFMTC, 9 August 1951.

The 6555th, Chapter II, Section 1

MATADOR and the Era of Winged Missiles

MATADOR Operations Through 1954

Though the Army's BUMPER launches at Cape Canaveral were followed by the first launch of a REDSTONE ballistic missile in late August 1953, aerodynamic or "winged" missiles dominated the Cape's launch schedule for most of the 1950s. That decade witnessed the introduction of the MATADOR, SNARK, BOMARC, NAVAHO and MACE aerodynamic missiles, among which the MATADOR, with over 280 launches to its credit, stood out as the most-launched missile of its era. The MATADOR was also the Cape's first full-fledged weapon system program and its initial deployment overseas included military launch crews trained by the 6555th Guided Missile Squadron at Cape Canaveral. Follow-on testing of the missile provided refinements in its performance as well as realistic training for several MATADOR squadrons under Tactical Air Command (TAC). As a direct descendant of the MATADOR, the MACE benefited from the "lessons learned" during MATADOR R&D testing in the 1950s. Ultimately, the MATADOR had a profound impact on the 6555th's organization, manpower and "blue suit" launch traditions.[1]

Under the terms of a missile contract, the contractor was responsible for development of a weapon system based on ARDC-approved technical requirements. Once a missile program reached the Cape, the Air Force Missile Test Center (AFMTC) was charged with acquiring and recording data to confirm those technical requirements were being met. Missile tests on the Eastern Test Range focused on missile performance first, but they soon provided an opportunity for military participation in launch operations. Like its predecessors at Eglin, Holloman and Patrick, the 6555th Guided Missile Wing was given the pivotal role of observing the contractor's operations and analyzing the results of each test. This function was designed to minimize additional validation launches, since military witnesses could confirm the contractor's compliance with basic test objectives during the R&D portion of a missile program. With regard to "operational suitability," AFMTC planned to acquire a minimum launch capability for all missiles that came its way, and it set about developing groups of trained military personnel to assemble, check out, prepare, launch and guide missiles assigned to AFMTC for testing purposes. For the MATADOR program, some of the 6555th Guided Missile Squadron's observers and reporters became the members of the initial MATADOR launch cadre, and they passed their training on to the first two operational MATADOR units -- the 1st and 69th Pilotless Bomber Squadrons (Light). This training included on-the-job factory training, missile assembly shop training, contractor and military instruction and simulated and real MATADOR launches.[2]

MATADOR operations became a large part of the 6555th's mission in the early 1950s, but the Wing had to consider its other agencies and tasks as well. The 4803rd Guided Missile Squadron, for example, had been launching LARKS from Cape Canaveral since October 1950. That mission continued under the 6556th Guided Missile Squadron with the launch of two more LARKS in June 1951. Detachment 1 also continued to support RAZON guided bomb tests at the Eglin Air Proving Ground through the end of July 1951, and its 30 officers and men were transferred to Patrick for reassignment to other duties within the 6555th Guided Missile Wing in August 1951. The 6555th Test Support Squadron was activated on 4 September 1951 to operate and maintain various types of "chase" planes and control aircraft being used to support the Wing's various guided missile projects. In addition to its LARK operations, the 6556th Guided Missile Squadron established a FALCON cadre at Holloman Air Force Base on 31 March 1952. It also organized a RASCAL cadre at Patrick on 16 June 1952 and sent it to Holloman. (Both cadres were transferred to the Holloman Air Development Center in early September 1952.) When the 6555th Guided Missile Wing was redesignated as a Group on 1 March 1953, most of its headquarters functions were dropped, and the 6556th Guided Missile Squadron and 6555th Test Support Squadron were discontinued. Nevertheless, the 6555th Guided Missile Wing continued to have many time-consuming tasks apart from MATADOR.³

FALCON MISSILE - 1952

RASCAL AIR-TO-SURFACE MISSILE - 1951

SHRIKE TEST MISSILE - 1952

SHRIKE IN-FLIGHT LAUNCH - 1950

From its inception the 6555th Guided Missile Wing had a multi-faceted mission. As Wing Commander, Colonel George M. McNeese was responsible for: 1) organizing, supervising and conducting guided missile tests assigned to AFMTC, 2) developing handling techniques and tactics, 3) training cadres for tactical missile units, and 4) submitting reports on missile research and development. Apart from the five officers and 25 airmen at Eglin, Colonel McNeese had 30 officers and 71 airmen working on various tasks at his Headquarters in June 1951. Fifty-two officers and 410 airmen were also assigned to the 6555th's two squadrons during this period. The 6555th Guided Missile Squadron, with 33 officers and 284 airmen, was in training to assist the Glenn L. Martin Company with its MATADOR launch program. The Squadron also planned to conduct operational suitability tests on the MATADOR and train the first two operational MATADOR squadrons (i.e., the 1st and 69th Pilotless Bomber

Squadrons), which were activated at Patrick in October 1951 and January 1952. The 6556th Guided Missile Squadron, commanded by Lieutenant Colonel Henry B. Sayler, had 19 officers and 126 airmen involved in LARK operations as a functional training exercise in anticipation of more advanced surface-to-air missile programs like the BOMARC. With the creation of the 6555th Test Support Squadron on September 4th and the activation of the 1st Pilotless Bomber Squadron on October 1st, the Wing had 119 officers, two warrant officers and 599 airmen assigned to its various operations by the end of October 1951. The Wing's strength increased rapidly after the 69th Pilotless Bomber Squadron's activation on 11 January 1952. By the end of June 1952, Colonel McNeese had the following resources:[4]

ORGANIZATION	OFFICERS	AIRMEN	TOTAL
WING HQ	32	132	164
6555 TSS	8	137	145
6555 GMS	28	244	272
6556 GMS	29	239	268
1 PBS	31	326	357
69 PBS	41	256	297
TOTAL	169	1334	1503

The MATADOR (B-61) program commanded most of the 6555th's attention during its first four years at Cape Canaveral, so it is only fair to begin our review of the winged missiles with the MATADOR. Between 20 June 1951 and 23 May 1952, 18 bright-red MATADOR "X" and "Y" experimental missiles were launched from Cape Canaveral by the Glenn L. Martin Company and the 6555th. All but one of the launches validated the MATADOR's zero-length launcher. Nine of the flights confirmed that the MATADOR's airframe was airworthy, and several of the later flights verified the usefulness of a control system prototype. The 6555th Guided Missile Squadron assisted the contractor in checking out and launching 16 of those missiles, including Number 547, which was the first B-61 prepared and launched successfully by an all-military crew on 7 December 1951. Thus, by the time the 1st and 69th Pilotless Bomber Squadrons were ready to begin training in 1952, the 6555th was prepared to provide that instruction.[5]

MATADOR LAUNCH - 18 Jul 1951

MATADOR ON ROADABLE LAUNCHER - 1953

MATADOR TRANSPORT TRAILER - 1953

The 6555th's MATADOR training program was divided into three phases. During the first phase, personnel assigned to propulsion and missile assembly received 13 days of individual training, and individuals assigned to missile guidance were given 43 days of instruction. During the second phase, individual technicians were gathered into three distinct types of teams (e.g., assembly, checkout or launch) to start working on a MATADOR missile. This phase normally took about six weeks, but lack of training missiles and ground equipment often conspired to make this phase of training longer. In the third phase, guidance, propulsion and assembly teams were joined together as crews. Crew training was expected to last 40 days, depending on the availability of training missiles. A final phase of training -- conducted by the squadrons themselves -- turned the crews into an operational squadron under its own staff officers and commander. During this final phase, the 6555th's instructors operated in an advisory capacity only.[6]

With regard to the training actually conducted in 1952, the 1st Pilotless Bomber Squadron began its individual training on 16 January 1952. Its team training in assembly, propulsion and controls started February 2nd and continued through March. (Crew training in those areas caught up with crew training in the guidance area in April.) During the crew phase, missiles were assembled and checked out using assembly line procedures. Technicians checked engine systems and engine run-ups followed. The launch crew took over and completed a simulated launch of the missile. (Initially, training missiles were provided by the 6555th, but the 1st Squadron received its first training missile -- Number 553 -- in June 1952.) Due to its later activation date and a lack of training equipment, the 69th training on airborne guidance and flight control equipment, RATO equipment and missile engine systems until April 1952. Members of the 69th's launching section faced even longer delays, and their individual training did not begin until early June. Both squadrons were "basically trained" by the end of 1952, but the lack of special squadron equipment and training launches stymied efforts to make either squadron operational by an early date. Problems with the MATADOR's performance also delayed the deployment of both squadrons.[7]

At this point, we need to take a closer look at the MATADOR and its ground and flight support equipment. Early test versions of the MATADOR were slightly more than 34 feet long and 23 feet from wing-tip to wing-tip, but they evolved into production models measuring 39.6 feet by 28.7 feet. In either configuration, the MATADOR could carry approximately 250 gallons of JP-3 jet fuel, and it weighed about as much as a jet fighter of comparable size. The missile's warhead compartment was designed to carry a 3,000-pound weapon, but it carried test equipment and ballast for the MATADOR's flights from Cape Canaveral. The airframe was designed to handle the combined thrust of the missile's RATO solid rocket booster and the Allison J-33-A-31 turbojet engine (i.e., 44,600 pounds of thrust). The wing and tail surfaces were aluminum alloy shells reinforced with honeycomb cores. Spoilers provided lateral control, and the MATADOR's horizontal tail was mounted atop the vertical stabilizer to minimize buffeting at high sub-sonic speeds. The missile's zero-length launcher was a 20-ton flat-bed trailer equipped with a cradle to support the MATADOR and elevate the missile to a launching angle of 18 degrees. A transport trailer was used to carry the MATADOR and its wing as two separate pieces to be assembled at the launch site.[8]

Two different guidance systems were under development for the MATADOR program in the early 1950s -- the SHANICLE and the MARC. The SHANICLE (Short Range Navigation Vehicle) system consisted of a microwave pulsed hyperbolic network based on principles applied in the common LORAN navigation system. The SHANICLE employed a pair of "master" and "slave" microwave ground stations to generate an azimuth for the missile's flight to the target. It used a second pair of master/slave stations to generate the distance to the target. The intersection of azimuth and distance hyperbolas defined the target. The master stations controlled timing, synchronized the microwave signals to the slave stations and (most importantly) transmitted guidance signals to the MATADOR as regular intervals during the flight. Once the missile reached its "terminal dive" point near the target area, a signal was sent to precess the MATADOR's vertical gyro, and this action sent the missile into a vertical dive toward the target.[9]

SCR-584 RADAR - Cape Canaveral, 1953

The MARC (MATADOR Automatic Radar Command) guidance system was an adaptation of the MSQ-1 radar system used to direct fighter-bombers during the Korean War. The MARC employed a modified SCR-584 ground radar to track an AN/APW-11 control beacon mounted in the MATADOR. Based on distance, direction, ground speed and altitude data received from the beacon, the MATADOR's position was computed in relation to the target and displayed continuously on an AN/MPS-9 plotting board. The radar controlled the MATADOR's flight via signals transmitted to the beacon control unit, and, once the missile reached the target area, a signal from the ground precessed the missile's vertical gyro and sent the MATADOR into a vertical dive toward its target.[10]

MSQ-1 CONSOLE - 1952

In mid-December 1950, the MARC was introduced as an alternative to the SHANICLE guidance system, but it soon became the front-runner in missile guidance tests at Cape Canaveral. Two MSQ-1 radars were transferred to the 6555th Guided Missile Squadron for the MATADOR program in September 1951, shortly after Lieutenant Colonel Richard W. Maffry assumed command of the Squadron. Another AN/MSQ-1 radar arrived from the Glenn L. Martin Company in March 1952, and it was set up at Jupiter Inlet, about 95 miles south of Patrick Air Force Base. During this period, technicians from the Rome Air Development Center trained the 6555th's MSQ Section in the operation and maintenance of the radars. On 4 April 1952, three of the 6555th's officers participated in the MARC's initial MATADOR flight, and they proved the value of the MARC by controlling the missile successfully over its entire 25-minute-long flight downrange. A simplified terminal dive system prototype was also introduced in 1952, and that system improved the MATADOR's response to terminal dive commands.[11]

In addition to the controls provided by SHANICLE or MARC stations on the ground, the MATADOR's experimental flights could be controlled from the air by a command radio system installed in an F-86 director aircraft. This command radio system let the controller adjust the MATADOR's throttle, rudder and control surfaces. The system also allowed him to: 1) override the missile's automatic control system to "dump" the missile and 2) override the MATADOR's fail-safe destruct system to save the missile. In addition to extending MATADOR flights and providing the contractor with more data on each missile test, the airborne command system offered safety advantages, since the director aircraft could be used to steer the missile away from populated areas, ships or other assets that might otherwise be left at risk if ground station signals faded. It should be noted, however, that the air support required for each MATADOR flight was rather extensive: in addition to the F-86 director aircraft, one B-29 (simulated missile) aircraft, one C-47 guidance synchronization aircraft, two B-17 airborne radar surveillance aircraft, one B-29 interference control aircraft and one C-47 range clearance aircraft were required. Those aircraft were maintained and operated by 6555th Test Support Squadron, until that unit was discontinued on 1 March 1953.[12]

ELECTRICAL PLOTTING BOARDS
Cape Canaveral, 1953

At the beginning of 1953, the 6555th Guided Missile Squadron anticipated at least 75 more MATADOR launches at the Cape, including 30 flights to determine the missile's operational suitability. The 6555th Test Support Squadron also expected to fly aircraft in 30 additional simulated MSQ-1 flight tests to: 1) confirm the MARC system's reliability and 2) to provide training for MSQ-1 operators and technicians in the 1st and 69th Pilotless Bomber Squadrons. When the 6555th Guided Missile Wing became a Group on 1 March 1953, the 6555th Test Support Squadron was discontinued, but the loss of the 6555th's "air arm" did not affect the ground aspects of the Group's MATADOR mission. The 6555th Guided Missile Group and its three remaining squadrons began using two of Patrick's newest buildings (Hangars "A" and "B") for missile assembly and checkout operations at the end of June 1953, and 23 more MATADORs were launched from the Cape in the last six months of 1953. As part of those operations, the 1st Pilotless Bomber Squadron concluded its training program by launching five B-61A MATADORS within a 22-hour period on December 15th. All five missiles were launched successfully, and four of them went into the designated impact area.[13]

COLONEL ALBERT G. FOOTE

MSQ-1 INSTALLATION - 1955

The MATADOR still had some technical problems to iron out, including a tendency to break up during the terminal dive phase when the missile cracked the sound barrier (e.g., at speeds from Mach .95 to Mach 1.15). Since vibration appeared to dampen out when the missile exceeded Mach 1.15, one possible solution to the problem was to "dive" the MATADOR from high altitude, pushing the missile through the sound barrier quickly. This solution was attempted on at least two occasions, but it did not solve the problem completely. This left the contractor's engineers with two other possible solutions: 1) reinforce the MATADOR's structure to withstand the strains of a "worst case" terminal dive or 2) reduce the missile's speed in the terminal dive so that it never exceeded Mach .95. By adding two hundred pounds of structural reinforcement to the MATADOR's wings and tail, Martin gained the advantages of both solutions. The reinforced structure handled terminal dive vibrations better, and the addition weight reduced terminal dive speeds to less than Mach .95. Seven of the 23 MATADORs launched in the last half of 1953 dealt with structural integrity and terminal dive problems. They also revealed an underlying problem with the missile's hydraulic control system, which was subsequently determined to be "an accumulation of minor maladjustments or malfunctions." The Wright Air Development Center worked out a satisfactory "fix" with the contractor, and operational suitability testing was completed in July 1954.[14]

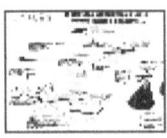
TECHNICAL OPERATIONS OF A DOWN RANGE STATION

HANGARS A AND B - Patrick AFB, 1953

In the meantime, the 6555th Guided Missile Group made a concerted effort to prepare the 1st and 69th Pilotless Bomber Squadrons for their reassignment to the Tactical Air Command (TAC) and their subsequent deployment to West Germany. Though chronic shortages in training missiles and other field training equipment had been redressed somewhat by December 1953, delays in the MATADOR program in 1952 placed both squadrons in an awkward position at the end of 1953: a considerable number of the 1st Pilotless Bomber Squadron's technicians had just returned from overseas assignments or were close to the end of their service commitments. With the 1st's departure set for early 1954, some of the 6555th Group's better-trained specialists had to be transferred to the 1st so it would have a full complement of personnel as close to the departure date as possible. By the end of December 1953, the 1st was 82 percent manned with overseas eligibles, and it had a solid nucleus of officers and non-commissioned officers supplemented by fully-trained resources from the 69th and the 6555th.[15]

The 1st and 69th were reassigned to TAC on 15 January 1954, but the 6555th Guided Missile Group continued to provide both squadrons with logistic and administrative support pending their overseas deployments. After the 1st Pilotless Bomber Squadron departed Patrick for Germany on March 9th, training in the 69th Pilotless Bomber Squadron intensified: under Lieutenant Colonel Maffry's command, the 69th had already launched three MATADORS in a highly successful field training operation on January 8th. In April, it fired 13 MATADORs in three other multiple-launch operations. By the end of June, the 69th had launched 30 missiles on extended flights (e.g., approximately 500 nautical miles in length), at night, during the day and in all kinds of weather. Its training completed, the 69th was relieved from AFMTC on 15 September 1954, and it departed for Germany.[16]

1ST AND 69TH PBS TRANSFER CEREMONY, 15 JANUARY 1954
Pictured Left To Right: Major General Richardson, Lt. Colonel Carroll from the 1st PBS, And Lt. Colonel Maffry from the 69th PBS.

Since TAC agreed to train all of its later MATADOR squadrons at TAC's own MATADOR school in Orlando, Florida, the 6555th Guided Missile Group was little more than a squadron when the 69th completed its field training in the summer of 1954. With no new pilotless bomber squadrons to train or support, most of the Group's staff were transferred to duties under AFMTC's Missile Operations Division. Under the command of Captain Edward B. Blount, the 6555th's Headquarters spent most of July and August liquidating its supply accounts, reassigning people and transferring property to other AFMTC units. The Group was discontinued on 7 September 1954, but the 6555th Guided Missile Squadron survived, and it was reassigned to AFMTC Headquarters on the same date. In October, the Squadron was reduced to a token force consisting of Captain Blount and four airmen, but AFMTC decided to restore the 6555th's MATADOR launch capability in December. Under Lieutenant Colonel Max R. Carey, the 6555th launched a MATADOR on 16 December 1954. Thirteen officers and 135 airmen were assigned to the 6555th by the end of December 1954.[17]

CAPTAIN EDWARD B. BLOUNT

The 6555th, Chapter II, Section 2

MATADOR and the Era of Winged Missiles

MATADOR and MACE Operations 1955-1963

In March 1955, the Glenn L. Martin Company phased out its testing crew, and all remaining MATADOR launches at Cape Canaveral were accomplished by military organizations. Those launches continued to pursue missile improvements as much as training requirements, and TAC's newest MATADOR unit-the 11th Tactical Missile Squadron-helped the 6555th Guided Missile Squadron test refinements in the AN/APW-11A beacon during launches required by the 11th's training program. Nine MATADORs were also launched by the 6555th during the first six months of 1955 to test a redeveloped version of the SHANICLE guidance system. The 11th launched 10 training missiles in June and July 1955, and the 6555th launched two MATADORs in September and a third missile in November 1955 to test the weapon's reliability on alert. [18]

The 6555th changed commanders several times over the next four years as the MATADOR mission continued. Major W. F. Heisler assumed command of the 6555th Guided Missile Squadron in May 1955, and he was succeeded by Major R. W. Cullen the following February. Major Cullen was promoted to lieutenant colonel during his tenure as Commander, which continued through early August 1958. Lieutenant Colonel John A. Simmons, Jr. took command subsequently and continued in that position until the Squadron became the 6555th Guided Missiles Group (Test and Evaluation) on 15 August 1959. Colonel Harry J. Halberstadt became the new commander following the Squadron's redesignation on August 15th, but he was succeeded by Colonel Henry H. Eichel on 21 December 1959, when the 6555th Group was reassigned from AFMTC to the Air Force Ballistic Missile Division (without any change in station) and redesignated the 6555th Test Wing (Development). [19]

MATADOR TRAINING MISSILE - 1954

The 6555th concluded MATADOR research and development testing at Cape Canaveral in 1956 as training launches continued. In the first six months of the year, 18 MATADORs were launched from the Cape. Twelve of them were launched to provide tactical training and to collect data on the missile's reliability and accuracy with the SHANICLE guidance system. Five of the launches were conducted mainly for training purposes, and one MATADOR was fired to evaluate the performance of a new missile launcher, the ASTRAL. A 19th missile was also launched from Patrick Air Force Base in the first public demonstration of the weapon system on May 20th, Armed Forces Day. Twelve more MATADORs were launched in the last half of 1956, including the 17th Tactical Missile Squadron's first training launch on August 29th. The 17th launched five more MATADORs by the end of September, and the 6555th Guided Missile Squadron concluded MATADOR R&D testing at the Cape with six MATADOR (SHANICLE) launches between September 20th and the end of November 1956. By the end of the year, one MSQ-1 guidance set was turned in to Base Supply for shipment back to the manufacturer (i.e., the Reeves Instrument Corporation), and SHANICLE base station equipment was turned over "in place" for use by the 17th Tactical Missile Squadron. On 12 February 1957, the TM-61 (MATADOR) Division was deleted from the Air Force Missile Test Center, and procedures were established which allowed AFMTC's Directorate of Range Operations to deal directly with TAC's newest MATADOR unit, the 588th Tactical Missile Group. For all practical purposes, the 6555th's involvement in the MATADOR program ceased at that time. [20]

MATADOR FIRED AT ARMED FORCES DAY CELEBRATION
20 May 1956

PUBLIC FIRING OF A MATADOR MISSILE AT PATRICK
AFB
20 May 1956

As a point of interest, MATADOR operations at the Cape continued under TAC for several more years. After several postponements in the training schedule, the 17th Tactical Missile Squadron resumed launch operations on 10 April 1957, and it completed its training program by launching four missiles by the middle of May 1957. The 17th was replaced by the 588th Tactical Missile Group, and the 588th conducted nine MATADOR launches between 29 August and 15 November 1957. (In an effort to improve the realism of this training, the 588th's people bivouacked in tents about three miles north of the MATADOR launch area.) Four more MATADORs were launched in March 1958, and the 588th concluded its training with six MATADOR launches in November and December 1958. The 588th was replaced by the 4504th Missile Training Wing in 1959. The 4504th launched 14 missiles between 7 January and 12 June 1959, and it launched 11 MATADORs between 25 August and 10 December 1959. Tactical Air Command's training units continued to launch MATADORs well into 1961. The last MATADOR was launched from Cape Canaveral on 1 June 1961. [21]

Well before the MATADOR program ended, the Glenn L. Martin Company introduced the MACE B to Cape Canaveral as a follow-on "cruise" missile program. The MACE B was an improved version of the MACE A, the MATADOR's immediate successor. The missile's lineage was apparent from its swept-wing, turbojet design: it was equipped with an Allison J33-A-41 engine, and, in its field configuration, it was launched from a "hard site" with the assistance of a RATO solid rocket booster weighing 2950 pounds. The missile was 44.3 feet long, 22.9 feet wing-tip to wing-tip, and its fuselage was 54 inches in diameter. The first MACE B (TM-76B) missile was launched from the Cape on 29 October 1959, and it met virtually all of its test objectives. The second MACE B launch, on December 4th, also demonstrated the value of the inertial guidance system and the missile's ground support equipment. [22]

As mentioned earlier, the 6555th Guided Missile Squadron went through some dramatic organizational changes in the last half of 1959, including the unit's redesignation as the 6555th Guided Missile Group (Test and Evaluation) on August 15th and the Group's reassignment and elevation to Test Wing status under the Air Force Ballistic Missile Division (AFBMD) on 21 December 1959. Despite the 6555th's reorientation to ballistic missiles, the Wing's MACE Operations Division (under the direction of Major Abbott L. Taylor) proceeded to develop a "blue suit" (all military) launch capability. Major Taylor's division had been in existence "in one form or another" since 1 July 1958, and the Division's key personnel completed factory training at Martin's Baltimore plant and participated in MACE B launches at Holloman Air Force Base, New Mexico before assisting Martin with the MACE B launches at Cape Canaveral in October and December 1959. [23]

Two MACE Bs were launched from a "soft site" on Complex 21 in February and March 1960, and an integrated military/contractor crew checked out and launched three more MACE Bs from the soft site by the end of June. The MACE B's hard site supported its first launch on 11 July 1960, and the military/contractor team launched three more missiles in September and October before the first two blue suit (all military) launches were conducted on 15 November and 16 December 1960. Following six more launches in March, April and June, the MACE B's final Performance Demonstration launch was completed on 21 June 1961. The 6555th's MACE Operations Division participated in all of those launches, and it completed its training supervision of TAC personnel assigned to MACE B operations at the Cape. The Division was phased out subsequently, and the MACE Weapons Branch (composed of five senior civil service engineers and 14 airmen) was established on 10 July 1961 to provide instrumentation support and engineering evaluation for 16 MACE Bs launched by TAC's 4504th Missile Training Wing. The MACE Weapons Branch was dissolved at the conclusion of the MACE Category III Systems Operational Testing and Evaluation (SOTE) program in April 1962. Its civil servants were transferred to the MINUTEMAN and ATLAS ballistic missile programs, and its airmen moved into positions with the ATLAS, TITAN and BLUE SCOUT programs. Tactical Air Command sponsored eight more MACE B missile launches at Cape Canaveral between 31 October 1962 and 18 July 1963. [24]

MACE LIFT-OFF FROM COMPLEX 21 "SOFT SITE"
11 February 1960

MACE HARDSITE PADS 21 AND 22 - January 1960

FLAME DEFLECTORS, PADS 21 AND 22 - January 1960

The 6555th, Chapter II, Section 3

MATADOR and the Era of Winged Missiles

LARK, BOMARC and SNARK Operations

Another thread winding through the 6555th's history was the unit's involvement with the LARK and BOMARC programs in the 1950s. As we noted in the previous chapter, the 3rd Guided Missiles Squadron launched three LARK surface-to-air missiles at Cape Canaveral in October and November 1950, and the 4803rd Guided Missile Squadron and the 6556th Guided Missile Squadron launched six more LARKs during the first half of 1951. LARK operations reached the operational suitability stage by the last half of 1951, but the "targets" were often nothing more than corner reflectors born aloft on balloons. Support requirements remained modest because the LARK was being used as a training foundation for the BOMARC program -- it was not a tactical missile program in its own right. By January 1952, the 6556th had 30 officers and 208 airmen, and it organized two missile teams to accelerate LARK training for its newly assigned personnel in February. Eight LARKs were launched during the first half of 1952 with mixed results, and the 6556th fired eight more LARKs before the Squadron was absorbed by the 6555th Guided Missile Squadron on 1 March 1953. The 6555th Guided Missile Squadron's LARK Branch launched seven missiles before terminating the training program on 8 July 1953.[25]

As the LARK program ended, attention shifted to the BOMARC, which was being developed as a tactical surface-to-air weapon system by the Boeing Aircraft Company and its sub-contractors (e.g., the University of Michigan, Westinghouse, Marquart Aviation Corporation and the Aerojet Corporation).** BOMARC operations began at AFMTC toward the end of June 1952, but the first missile arrived later than anticipated, and equipment shortages conspired to delay the first launch until 10 September 1952. Other launches were also slow in coming, due to the contractor's insistence that only one missile be tested at AFMTC at any one time. In effect, all flight data reduction and analysis had be completed at the Boeing plant in Seattle, Washington before the next missile was fired at Cape Canaveral. The second BOMARC was launched from Cape Canaveral on 23 January 1953, and the third BOMARC flight followed nearly five months later, on June 10th. Two more missiles were launched in the summer of 1953, but only three BOMARCs were launched from the Cape in 1954.[26]

Unlike the LARK program, the BOMARC test program at the Cape was essentially a contractor-led operation. The 6555th's people were not responsible for any BOMARC launches, but six airmen from the 6555th's 20-man BOMARC Section were assigned to help Boeing with electronic equipment maintenance tasks in late March 1953, and nine other airmen assisted the University of Michigan with its BOMARC activities at the Cape. The Air Force Missile Test Center provided range support and test facilities at the Cape, and AFMTC's safety agencies were responsible for insuring that safety requirements for the 15,000-pound, 47-foot-long missile were "stringently enforced." In relation to other aerodynamic missile programs at the Cape, the BOMARC continued to move ahead slowly: by the middle of 1956, only eight propulsion test vehicles, nine ramjet test vehicles and five guidance test vehicles had been launched. Two tactical prototype BOMARCs were launched against a QB-17 target drone in October and November 1956, but the 6555th's people only played a supporting role in those tests and later contractor-led operations.27

ACID FUELING OPERATION, BOMARC MISSILE
September 1952

BOMARC - August 1952

BOMARC FLIGHT SEQUENCE - 1952

BOMARC IN LAUNCH POSITION - 1956

Twenty-five more BOMARCs were launched from the Cape before ARDC announced plans in September 1958 to transfer the BOMARC program from Cape Canaveral to the Air Proving Ground Center's test site at Santa Rosa Island near Fort Walton Beach, Florida. This move was designed to focus AFMTC's efforts on ballistic missile test programs, but it also confirmed the fact that the Range had been selected for the BOMARC primarily because of its instrumentation capabilities, not because the 6555th had established a blue suit launch capability in other aerodynamic missiles at the Cape. In any event, one officer and 27 airmen were released from the 6555th Guided Missile Squadron and transferred to Hurlburt Field, Florida to support the BOMARC program in the fall of 1958. Boeing conducted its last Cape Canaveral launch of the BOMARC on 15 April 1960.28

BOMARC LAUNCH - 21 August 1958

Two more aerodynamic missile programs -- SNARK and NAVAHO -- need to be reviewed before we move on to the 6555th's involvement in ballistic missile programs at the Cape. Though neither program was particularly successful, the subsonic SNARK and the supersonic NAVAHO were a serious reflection of their times and an important part of the Cape's history: they were undertaken to give the United States an intercontinental cruise missile capability when confidence in an intercontinental ballistic missile capability remained slim.

The earlier of the two programs, SNARK, was initiated by the Northrop Aircraft Company in March 1946 to provide the Air Force with a turbojet-powered, subsonic, guided missile capable of carrying a 7,000-pound warhead up to 5,500 nautical miles. Northrop planned to meet a 600 mile-per-hour speed requirement initially, but the Air Force concluded that a faster missile with a supersonic dash capability would be needed by the time the SNARK was expected to go into production (i.e., around 1954). By 1950, Northrop was hard at work on an improved SNARK that could cruise at approximately .94 Mach. A dummy version of the missile was released from a track launcher at Holloman Air Force Base for the first time on 21 December 1950. Over the next eight months, Northrop conducted nine N-25 research vehicle flights at Holloman to test aerodynamic and guidance characteristics that would be incorporated in the longer, heavier N-69 test missile. The contractor planned to transfer the SNARK test program to AFMTC by the end of 1951, but the Cape lacked adequate missile assembly and hangar space, and additional facilities had to be built before the program moved to AFMTC in the spring of 1952. Ten SNARKs were launched at Holloman between the end of August 1951 and the end of March 1952, and Northrop continued work on the missile and its guidance systems at the company's plant in Hawthorne, California.[29]

ARRIVAL OF SNARK MISSILE AT PATRICK

June 1952

As an intercontinental weapon system, the SNARK would ultimately fall to the Strategic Air Command (SAC), but, in accordance with its mission, the 6555th Guided Missile Wing was directed to develop its own blue suit launch capability well in advance of SAC's units. Toward that end, the Wing received its first SNARK training missile (e.g., an N-25 research vehicle) in late May 1952, and the 6556th Guided Missile Squadron activated a SNARK cadre at AFMTC on June 16th. Under the command of Major Richard E. Eliason, the SNARK cadre had eight officers and 48 airmen at Northrop's plant for factory training by the end of June 1952. Over the next six months, the Wing hoped to have 11 officers and 64 airmen qualified to do unit training in guidance control, tape preparation and missile inspections.30

As the program stood at the beginning of 1953, the SNARK (B-62) production missile would be five feet in diameter and approximately 74 feet long. Its sharply-swept wings, projecting straight out from the fuselage, would span 42.5 feet. The missile's YJ71-A-3 Allison turbojet engine was expected to provide main power up to Mach .94, but an afterburner was under development to give the SNARK a supersonic dash capability. Two solid rocket boosters -- each rated at 105,000 pounds of thrust -- would launch the SNARK from its zero-length launcher and accelerate the missile to about 365 miles per hour. Though the SNARK was expected to weigh over 58,000 pounds on the ground, the rocket boosters would be released once the missile was airborne, and the SNARK's cruising weight would be less than 47,000 pounds.31

EARLY SNARK MISSILE - December 1952

After Northrop moved the SNARK program to Cape Canaveral, it boosted three dummy SNARK missiles from the SNARK's new zero-length launcher on 29 August, 1 October, and 30 October 1952. The results were satisfactory, and four radio-controlled N-25 SNARK launches followed on 26 November and 12 December 1952, and 6 February and 10 March 1953. All four N-25 flights were successful, and preparations got underway to launch the first N-69 SNARK test missile on 6 August 1953. Based on the SNARK's successes at Holloman and Cape Canaveral in 1952 and the first half of 1953, SAC planned to activate a 105-man cadre for its 1st Pilotless Bomber Squadron (Strategic) in January 1955. The 6555th's SNARK B-62 Operations Section (formerly the SNARK cadre) already had 76 officers and airmen in training to form the nucleus of a blue suit launch organization, and there was every indication in the spring of 1953 that those men would be involved in SNARK launch operations in the not-too-distant future.32

SNARK DUMMY MISSILE

Unfortunately, the SNARK suffered almost continuous set-backs after the N-69 made its debut at the Cape in the summer of 1953. Part of the trouble centered on the new missile's engine: though the N-25s had been powered by reliable J33A-31 or J35-A-23 Allison engines, the N-69's YJ-71 power plant malfunctioned repeatedly. Quality control became a problem at the Hawthorne plant, as evidenced by the incidence of missing parts and damaged components. In the last half of 1953, rework orders were posted against almost every SNARK sent to AFMTC for testing. New radio control equipment was unavailable, forcing further delays. The Northrop Field Test Crew managed to launch two SNARKs on 6 August and 15 October 1953, but the first SNARK crashed 15 seconds into the flight after its drag parachute deployed prematurely, and the second missile was destroyed after it became uncontrollable about five minutes into the flight. Modification orders continued to pour back to Northrop as a result of those failures, and the 6555th's SNARK section provided the contractor with more than 32,000 man-hours of direct support during the last half of 1953. In 1954, Northrop decided to drop the Allison engine in favor of the Pratt Whitney J-57 engine. The company also had the solid rocket boosters upgraded to 130,000 pounds of thrust each. Though the SNARK's diameter and wing span remained virtually unchanged, its length was eventually shortened to 67.2 feet. The SNARK's gross weight, minus boosters, increased to 49,000 pounds.33

SNARK PRE-LAUNCH - Cape Canaveral, 1954

LAST MINUTE CHECK BEFORE SNARK ENGINE RUN-UP AND LAUNCH

The Northrop Field Test Crew launched 11 "recoverable" N-69A and N-69B model SNARKs in 1954 and 1955. None of those missiles were recovered successfully, but the Crew acquired considerable data from the flights. Northrop also gathered aerodynamic data from Model N-69C missiles used in terminal dive testing, which started in February 1955. Unfortunately, not all of this information was good news: though the N-69C's first flight on February 10th was excellent, data gathered on later flights of the N-69C and the first two flights of the N-69D forced Northrop to suspend all SNARK launches in February 1956 to correct "the unreliability of certain system components." Following about four months of trouble-shooting, the ban on SNARK flights was lifted. N-69C flights resumed in July, and they continued until delivery system tests were completed in October 1956.34

SNARK guidance test flights were halted as part of the overall suspension in February 1956, but they resumed with the launch of a third N-69D missile on 13 September 1956. Fourteen more N-69D missiles were launched over the next 11 months to evaluate the MARK I inertial guidance system, and most of those flights met all of their test objectives. Encouraged by the results, Northrop pushed ahead with the military demonstration phase of the program in June 1957. After a failed debut on June 20th, the N-69E operational prototype responded well during most of its flight on 16 August 1957. Another N-69E made a routine flight on September 19th, and two N-69Es made the first two SNARK flights to Ascension on 31 October and 5 December 1957. Those missiles were launched by the Northrop Field Test Crew, but two N-69Ds were launched by all-military crews on October 1st and November 20th. Blue suit launches of the N-69E were just around the corner. After launching five more operational prototypes between January 25th and May 6th, the Northrop Field Test Crew launched its last N-69E missile on 28 May 1958. On June 27th, SAC's 556th Strategic Missile Squadron launched its first SNARK (an N-69E) under the supervision of the 6555th Guided Missiles Squadron. The launch was also the first in a series of flights for the SNARK Employment and Suitability Test (E and ST) program.35

 *F-89 CHASE PLANE AND SNARK ON TEST FLIGHT - 1956*

It should come as no surprise that the SNARK's operational suitability flights were interwoven with its R&D flights. This policy was first established at AFMTC for the MATADOR program in the mid-1950s, and it was reiterated by the AFMTC Commander in February 1957. It was also supported by the Air Proving Ground Command, SAC and ARDC in a test requirement conference held at Patrick on 5 and 6 March 1957. All agencies agreed that approximately 95 percent of the SNARK's E and ST requirements could be met by observing N-69E R&D tests and reviewing the data obtained on those flights -- as long as blue suit launch crews were involved in the operations. Northrop planned its first N-69E launch for June, so the 6555th quickly dispatched one officer and 21 airmen to Northrop's Hawthorne plant to receive additional specialized training in the spring of 1957. After that contingent returned to the Cape in June, it passed its knowledge on to other personnel in the 6555th. Plans for SNARK crew training and squadron-level operational testing were completed a few weeks later, and the 6555th accomplished its first blue suit SNARK launches on 1 October and 20 November 1957.36

The 556th Strategic Missile Squadron was activated under the command of Lieutenant Colonel Richard W. Beck at Patrick on 15 December 1957. The 556th was assigned to SAC, but it started its on-the-job training under the direction of the 6555th Guided Missiles Squadron in January 1958. Some of the 556th's men participated in an "over-the-shoulder" training exercise with the Northrop Field Test Crew in March, and the Squadron's first simulated launch training was conducted on April 4th. The Air Force Missile Test Center picked up responsibility for SNARK operational evaluation testing on 14 May 1958 (i.e., two weeks before the Northrop Field Test Crew's last launch), and the 6555th supervised the 556th's first launch on 27 June 1958. The 556th also launched two N-69Ds in November and December 1958.37

Five more SNARKs were launched by the 6555th in the last half of 1958, but the 556th's crew training program was shortened dramatically after the Air Force decided to limit the SNARK's deployment to just one operational squadron. On 1 January 1959, SAC activated the 702nd Strategic Missile Wing (ICM-SNARK) at Presque Isle, Maine, and it assigned the 556th to the 702nd in April. Eighty SAC personnel were sent to AFMTC in the spring of 1959 for crew training, and the 556th participated in three production model (SM-62) SNARK launches before it departed for Maine on 7 July 1959. Though the 556th was inactivated on 15 July 1959 and absorbed by the 702nd, 188 additional SNARK missilemen were trained under the 6555th Guided Missiles Squadron's supervision by the end of December 1959.38

Five SNARKs were launched in the last half of 1959, bringing the number of flights to 86 since the program's inception. Unfortunately, the SNARK continued to display performance problems. An overall review of the R&D effort toward the end of 1959 concluded that, once airborne, the SNARK only had one chance in six of hitting the target area. The Air Force Missile Test Center recommended cancellation of the program, but Air Force Headquarters decided to continue the R&D program at the rate of about one launch per month through 1960. In addition to normal range support, AFMTC agreed to provide airmen from the 6555th to support blockhouse telemetry operations and provide engineering evaluation on the next three SNARK flights. Though Northrop's Field Test Crew was gone, the contractor still had a considerable number of technical personnel at the Cape. Northrop was directed to maintain 125 employees to meet its technical responsibilities as missile contractor on the final test flights. Eleven more SNARKs were launched in 1960 before the test program was closed out.39

 *SNARK LAUNCH - 1960*

The 702nd Strategic Missile Wing placed its first SNARK on alert at Presque Isle on 18 March 1960, and three more missiles were added to the Wing's alert force within a few months. Despite those encouraging signs, the 702nd was not declared "operational" until 28 February 1961. One month later, President John F. Kennedy declared the SNARK "obsolete and of marginal military value," and SAC inactivated the 702nd on 25 June 1961. In retrospect, the SNARK was an abysmal failure as a weapon system, but it gave SAC considerable experience in preparing, training and deploying other strategic guided missile cadres in later years. The same could be said for the missile's value to the 6555th, in the context of that unit's mission at Cape Canaveral.40

The 6555th, Chapter II, Section 4

MATADOR and the Era of Winged Missiles

The NAVAHO Program

No review of the winged missile era would be complete without some mention of the NAVAHO program. The NAVAHO was a very ambitious effort, but it was even less successful than the SNARK. Only two of three planned versions (e.g., the X-10 and the XSM-64) were ever launched at the Cape, and the program was cancelled by Air Force Headquarters in July 1957. Left-overs from the program were launched as part of the "Fly Five" project and Project RISE (i.e., Research in the Supersonic Environment), but neither effort was successful. Even if the NAVAHO had been successful as a weapon system, it would have been eclipsed by the ATLAS, TITAN and MINUTEMAN ballistic missiles in the early 1960s. Those criticisms aside, the NAVAHO proved that good things can come out of bad programs: the missile's inertial guidance system found its way successfully into nuclear-powered submarines, Navy attack aircraft, and the HOUND DOG and MINUTEMAN missiles. Though the 6555th had no involvement in the program, the NAVAHO's impact on AFMTC and Cape Canaveral must be mentioned to put other missile activities -- including ballistic missile programs -- in perspective.[41]

NAVAHO X-10 ON SKID STRIP - Cape Canaveral, 1956

NAVAHO X-10 AND RECOVERY CREW - 1956

NAVAHO XSM-64 AND BOOSTER MATED FOR LAUNCH - 1956

The goal of the NAVAHO program was to produce a surface-to-surface guided missile capable of carrying an atomic warhead 5,500 nautical miles at a speed of at least Mach 2.75 with sufficient accuracy to insure that at least 50 percent of all missiles struck within 1,500 feet of the target. The North American Aviation Company was the prime contractor for the missile, but the Wright Aeronautical Company had a contract to develop the ramjet engines that would be used on the XSM-64 (Phase Two) test vehicle and, presumably, the production (Phase Three) missile. During Phase One of the program,

Westinghouse provided North American with the 10,900-pound thrust J-40-1 turbojet engines that were used in pairs on single-staged, recoverable X-10 test vehicles. North American also procured pairs of 120,000-pound thrust rocket engines from its Rocketdyne division to power the booster stage of its two-staged XSM-64 test vehicle during Phase Two. A series of fifteen X-10 flights were conducted at Edwards Air Force Base, California as part of Phase One in 1953 and 1954. North American also began operating a small field office at Patrick Air Force Base in 1953 to coordinate support efforts for the program, including the construction of two missile assembly buildings, a vertical launch facility for the XSM-64 and a 200 x 10,000-foot landing strip on Cape Canaveral for the X-10 vehicle.[42]

NAVAHO ERECTING PEDESTAL AT 75 DEGREES

NAVAHO IN VERTICAL POSITION ON LAUNCH STAND
Cape Canaveral, 1957

The X-10's first flight was scheduled to be launched from the Cape in the summer of 1955, so North American tripled its field office staff from 22 to 77 people in 1954. It also began installing equipment in the guidance laboratory, the blockhouse and the NAVAHO'S flight control building even before construction of those facilities was completed. The first X-10 was launched from the Cape on 19 August 1955, and the NAVAHO quickly replaced the MATADOR as the Range's principal user (though only for the short term). Support facilities were completed in the last half of 1955, and seven more X-10s were launched from the Cape over the next twelve months. By the middle of 1956, North American had 605 people working on the NAVAHO program at Cape Canaveral and Patrick.[43]

Five more X-10 flights were completed in the last half of 1956, but problems with an auxiliary power unit held up the XSM-64's first launch until 6 November 1956. After six months of delays, the XSM-64's debut on November 6th was not encouraging: the pitch gyro failed 10 seconds after lift-off, and the missile and its booster broke up and exploded 26 seconds into the flight. Three more XSM-64s were launched over the next seven months with depressing, if not equally dismal, results. On 22 March 1957, the first of those missiles impacted 25 nautical miles downrange after the booster's engine shut down prematurely. The next missile fell back on the launch pad on April 25th after rising only four feet. (The subsequent explosion and fire did considerable damage to the pad.) The last of the three was launched on 26 June 1957. It performed well until the ramjets failed to operate after booster separation, and the missile impacted about 42 miles downrange.[44]

NAVAHO WRECKAGE NEAR PAD - 25 April 1957

In addition to those failures, the first in a series of 1,500-mile-long auto-navigator test flights was attempted 10 times in the first three months of 1957 without a single launch. The only bright spots in the program seemed to be some static tests of the NAVAHO's booster rockets and North American's isolation of problem areas revealed in the first four XSM-64 flights. Unfortunately for North American, NAVAHO was already doomed. In a message dated 12 July 1957, Air Force Headquarters terminated the NAVAHO's development. Because the auto-navigator showed promise, the Air Force authorized five auto-navigator flights and one radio command flight with no landing capability. The first of those flights was conducted on August 12th, and it suggested that the auto-navigator functioned properly -- at least until the left ramjet blew out about seven and a half minutes into the flight. The second flight, on September 18th, was even more successful, but the missile had to be destroyed after it entered a slow right turn about 450 miles downrange. Two other flights were less successful, but one NAVAHO managed to autonavigate as far as the Range's station on Mayaguez (about 1075 miles downrange) before its ramjets failed.[45]

SIDE VIEW OF NAVAHO XSM-64 ON LAUNCH STAND
Cape Canaveral, 1957

NAVAHO XSM-64 LAUNCH - 26 June 1957

The seven remaining XSM-64s were designated for supersonic research to support the B-70 bomber and long-range interceptor programs, but only two were ever launched from the Cape. The first of them reached Mach 3.1 at an altitude 63,000 feet on 11 September 1958, but its ramjets failed to ignite and the missile crashed 82 miles downrange. The other XSM-64 achieved Mach 3.0 at 77,000 feet before breaking up 60 seconds into its flight on 18 November 1958. The B-70 Weapons System Project Office urged termination of the RISE program as soon as possible, and no more XSM-64s were launched.[46]

As North American closed out the NAVAHO program, three X-10s were selected as support drones for BOMARC missile tests in late 1958 and early 1959. Two X-10 drones supported BOMARC launches successfully on 24 September and 13 November 1958, but both X-10s burned after running off the end of the Skid Strip at the end of their missions. The last X-10 was launched on 26 January 1959 with no apparent problems, but it self-destructed and crashed approximately 57 miles downrange after experiencing a power failure. It was the NAVAHO's final flight from Cape Canaveral.[47]

 *X-10 DRONE FLIGHT FROM SKID STRIP - 24 September 1958*

Like the SNARK, the NAVAHO had been an overly ambitious attempt to find the practical limits of the winged missile as a weapon system. Some good things came out of the program, but critics maintain (with some justification) that too much effort was expended on the SNARK and NAVAHO, and that the money spent on aerodynamic missile programs would have been better spent elsewhere. The SNARK certainly lingered far too long, but what might have happened to the NAVAHO if ballistic missiles had proved impractical? The ATLAS and TITAN were not assured success when their requirements were laid down in the 1950s. Who could say -- in 1953 -- whether the ATLAS or the NAVAHO would be the better missile? (Admittedly, future prospects looked better for the ATLAS and worse for the NAVAHO after 1956.) In the spirit of the times, aerodynamic and ballistic missile programs were pursued to provide a margin of safety against the failure of either type of missile. Air Force planners in the early 1950s did not have the luxury of our hindsight to guide their future and our past. The apparent anachronism of winged and ballistic missiles being launched within days of each other at Cape Canaveral can only be explained in that context. It is also true that there is normally some overlap between old and new technologies whenever a culture is about to make a great leap forward. The era of winged missiles helped shape the overlapping age of modern weapon systems, as evidenced by the transfer of guidance systems and other components to intercontinental ballistic missiles, cruise missiles, and tactical air-to-air missiles in the 1960s. In that respect, very little of the effort invested in the MATADOR, MACE, BOMARC, SNARK and NAVAHO was completely wasted.

The 6555th

Chapter Two Footnotes

6555th Test Support Squadron
Ten officers and 129 airmen under the command of Lieutenant Colonel John C. Reardon were present for duty in the new squadron on September 5th.

FALCON
The FALCON was a fighter-launched, supersonic, air-to-air missile with a range of about four miles. Weighing 122 pounds and measuring only 77.8 inches long, the missile carried a small explosive warhead activated with a contact fuse. The FALCON was developed through a series of prototypes (e.g., models "A" through "F"), and FALCON model "C" and "D" missiles were fired against bomber drones at Holloman in 1952. Captain Wilbur R. Lindsey, Jr., one other officer and 13 airmen from the 6556th reported to Holloman in early April 1952 to support the Hughes Aircraft Company with its FALCON. Two of Lindsey's airmen conducted telemetry operations for Hughes in June 1952, and eight other airmen were in training at Hughes' plant in California during the same period.

RASCAL
The RASCAL was a 32-foot-long, air-to-surface guided missile designed for all-weather use in medium and heavy bomber operations against strategic targets. The RASCAL was developed by Bell Aircraft Corporation under an Air Materiel Command contract as a supersonic cruise missile. A 2/3 scale version of the RASCAL called "SHRIKE" was tested at Holloman in 1951 and 1952 to evaluate the aerodynamics and launching characteristics of the RASCAL system. Though there was some thought given to transferring the RASCAL program to AFMTC in 1952, Headquarters ARDC decided to keep the RASCAL at Holloman along with shorter-ranged missile programs.

6556th Guided Missile Squadron and 6555th Test Support Squadron
The 6556th's people were absorbed by the 6555th Guided Missile Squadron. The Test Support Squadron's people were transferred to AFMTC's Air Support Squadron under the 6550th Air Base Group. The Air Support Squadron continued to support the MATADOR and other missile programs.

6556th Guided Missile Squadron
As mentioned earlier, the 6556th would be absorbed by the 6555th Guided Missile Squadron in March 1953, so BOMARC proved to be the limit of the 6556th's ambitions.

1st Pilotless Bomber Squadron
The 1st Pilotless Bomber Squadron was commanded by Lieutenant Colonel James Giannatti initially, but Lieutenant Colonel Louis O. Carroll assumed command on 19 November 1951. By the end of December 1951, Carroll's squadron consisted of 17 officers and 73 airmen, but tents had to be set up

near two of Patrick's barracks in December to shelter 174 additional airmen who reported to the Squadron in mid-January 1952.

69th Pilotless Bomber Squadron
Lieutenant Colonel George T. Walker assumed command of the 69th in January, and 41 officers and 256 airmen were assigned to his squadron by the end of June 1952.

Number 547
Personnel from the 6555th Guided Missile Squadron assembled and disassembled Number 547 several times before the launch to make the most of their training experience. Checks of the controls, guidance and telemetry system were also done repeatedly. Martin representatives stood by as consultants on December 7th and provided the test equipment for the launch. Lift-off and flight were normal, but the missile did not respond properly to guidance signals, and it finally went out of control and fell into the Atlantic 15 minutes and 20 seconds after launch. The flight covered a distance of 105 miles.

"basically trained"
This training prepared both squadrons to receive missile components, assemble them into a complete missile (minus warhead and RATO), perform system functional checks, set the missile on its zero-length launcher, attach the MATADOR's wing, warhead and RATO, insert targeting information, make final checks and launch the missile. Guidance personnel in both squadrons were trained to operate the MSQ-1 radar and control the missile along its flight path to the target. In addition to the training at AFMTC, both squadrons sent officers and airmen to Lowry Air Force Base, Colorado for guidance training, and to Chanute Air Force Illinois for propulsion training.

MARC
The decision to develop the MARC was based on the contractor's ability to carry out parallel development of both guidance systems without delaying the delivery of an operational MATADOR weapon system. The MARC was desirable because of its potential as a standardized guidance and control system for pilotless missiles and fighter-bombers.

Lieutenant Colonel Richard W. Maffry
Maffry assumed command on September 4th, when the Squadron's former commander, Lieutenant Colonel John C. Reardon, moved over to take command of the 6555th Test Support Squadron on the same date. Major John A. Evans succeeded Lieutenant Colonel Reardon as Test Support Squadron Commander on 12 May 1952.

fail-safe destruct system
The missile carried a positive destruct system which allowed controllers to destroy the missile on command, but the MATADOR also carried a fail-safe system which destroyed the missile automatically upon interruption of control signals for any period longer than 45 seconds. Following interruption of control signals, the missile continued on course for 15 seconds before executing a left turn. If signals were not restored within the next 30 seconds, the missile destroyed itself automatically. On the other

hand, if the missile was flying properly but went into a left turn because of an interruption in the ground signal, the controller in the director aircraft could switch on his radio command system and prevent the missile from destroying itself prematurely.

6555th Guided Missile Wing became a Group
The 6555th Guided Missile Squadron was discontinued on 1 March 1953 as well, but its resources were transferred to the 6555th Guided Missile Squadron. The Group still had 97 officers and 1038 airmen assigned to its units at the end of June 1953. Colonel Albert G. Foote became the Group Commander on 1 March 1953, succeeding Colonel Jack S. DeWitt, who had been the 6555th's Wing Commander since Colonel McNeese's departure for a new assignment on 16 July 1952. Foote was succeeded by Lieutenant Colonel Henry B. Sayler on 6 June 1953.

MATADOR school
One of TAC's MATADOR mobile training detachments was attached to the 6555th Guided Missile Squadron in 1954, and it was sent to Orlando Air Force Base in late November 1954 to begin training new MATADOR squadrons.

Captain Edward B. Blount
Captain Edward B. Blount commanded the 6555th Guided Missile Group from 22 June 1954 until the unit was discontinued, whereupon he assumed command of the 6555th Guided Missile Squadron. Lieutenant Colonel Carey assumed command of the Squadron in early December 1954.

11th Tactical Missile Squadron
The 11th had launched its first MATADOR on February 21st, and it launched three more in March, April and May 1955.

ASTRAL
The ASTRAL (Assembly Transport and Launch) launcher was designed by the Martin Company as both a transporter and launcher, thereby eliminating the need for two separate pieces of equipment. Built of tubular steel tied together with steel cables, the ASTRAL was lighter, less expensive and easier to operate than the older zero-length launcher. The 6555th tested the ASTRAL in a MATADOR launch on 2 May 1956. The new equipment functioned well and "incurred no incidental damage." Plans called for ASTRAL road tests and launch demonstrations in Europe in the last half of 1956.

first public demonstration
The missile used in the public demonstration had been used in the "ready storage" program, initiated at the Cape in October 1955. Under that program, the missile had been kept under a tarpaulin out of doors and monitored to see how long its systems remained functional. Weekly verification checks were conducted over the next six months, and, despite the highly corrosive environment of the Central Florida coast, the missile flew properly for the public on May 20th.

MACE B

The MACE A and MACE B looked alike, but their guidance systems were different. The MACE B used an inertial guidance system manufactured by the A.C. Spark Plug Company. The MACE A used the ATRAN/Automatic Terrain Recognition and Navigation System developed by the Goodyear Aircraft Corporation. As far back as 1952, Goodyear had been working on the ATRAN system for a 1,000-mile version of the MATADOR planned by Martin. Though the proposed vehicle was never fielded as a MATADOR, it became the MACE A missile, which carried its entire guidance system internally and did not have to rely on ground stations to guide it to the target. To accomplish this, the ATRAN system matched the missile's radarscope presentation of the immediate terrain with a previously obtained radar picture (simulated by radar-photo reconnaissance) and adjusted the missile's control surfaces to keep the missile on course into the target. The value of this technology was proven dramatically during the DESERT STORM campaign of 1991, in which air- and sea-launched cruise missiles played a decisive role in the early hours of the war in the Persian Gulf.

blue suit (all military) launches
The blue suit launch crews included four officers and 28 airmen from TAC and six airmen from Air Training Command. They were attached to the MACE Operations Division and integrated into launch crews as cadres for their parent commands' MACE B operations and training programs.

two missile teams
Team #1, consisting of three officers and 33 airmen, was divided into a propulsion section and a guidance section. The Propulsion Section was given classroom and shop training in boosters, motors, destructors and fueling systems. The Guidance Section was divided into sub-sections and given training in: 1) target seeker and attitude controls, 2) fire control and 3) receivers. Team #2, composed of three officers and 36 airmen, was responsible for checking out and launching the missiles prepared by Team #1.

BOMARC
In November 1949, the Air Force asked Boeing and the University of Michigan to make a feasibility study of a surface-to-air guided missile to supplement the nation's air defense forces. The contractors' joint study subsequently proposed a BOMARC (Boeing and University of Michigan Aeronautical Research Center) missile system tied to a network of searching and tracking radars. Under the BOMARC concept of operations, the tracking radars passed targeting information to a computer/evaluator system which, in turn, assigned BOMARCs to individual targets and fed the missiles guidance information to intercept their targets. Jet interceptors would also be integrated into the system to allow a joint air defense operation, a "missile only" operation or a "manned interceptor only" operation. The BOMARC's design and development phase was started in December 1950, and contractor compliance tests were underway in 1953. Boeing was awarded the contract for the missile, but sub-contracts went to the University of Michigan (for modifications to the computer/evaluator system), to Westinghouse (for target seekers), to Marquart (for the ramjet engines), and to Aerojet (for the missile's liquid rocket booster).

safety requirements
Unlike the MATADOR's solid propellant RATO system, the BOMARC's liquid rocket motor had to be

fueled with white fuming nitric acid and analine-furfuryl alcohol in two separate, potentially dangerous operations. A portable shower was erected at the launch pad, and a decontamination truck and crew stood by throughout both fueling sequences as a mandatory safety measure. A third fueling operation -- involving JP-3 jet fuel -- was similar to MATADOR pumping procedures, except for special precautions required by the presence of the missile's loaded liquid rocket. Like other JP-3 pumping operations, the BOMARC JP-3 fueling sequence was monitored by a Cardox (carbon dioxide) fire truck and crew. The decontamination crew, fire trucks, an ambulance and a doctor remained at the launch pad during all fueling operations.

target drone
The 3215th Drone Squadron from Eglin's Air Proving Ground Center provided the target drones for the BOMARC IM-99A test program. On 5 December 1958, the Squadron was discontinued, but it was succeeded by the 3205th Drone Group, Detachment #1, which continued flying drone targets for BOMARC tests well into 1959. Once the IM-99A portion of the program was completed, drones were no longer required. Detachment #1 departed for Eglin on 8 June 1959.

track launcher
The test vehicle was launched from a rocket sled mounted on a 3,300- foot length of railroad track. As the sled raced down the first 1,500 feet of track, it released the SNARK at approximately 350 miles per hour. Once the missile was airborne, the sled braked itself by means of a scoop, which plunged into a water trough located between the rails. The sled was powered by three 3-DS solid propellant motors rated at 47,000 pounds of thrust apiece.

N-25 research vehicle
The N-25 research vehicles were equipped with landing skids so the vehicles could be used on several flights. The N-25s were launched by rocket sled, and radio-controlled by missile pilots flying in director aircraft.

guidance systems
The SNARK had three guidance systems: 1) an APN-66 radar, 2) an ACN unit, and 3) an inertial terminal guidance system. Following launch, the SNARK was guided to a point in space by its APN-66 Doppler radar system. After a star reference was acquired, navigational control passed to the SNARK's Automatic Celestial Navigation (ACN) unit, which controlled the mid-course portion of the flight by comparing known star coordinates with the missile's pre-programmed flight plan. The inertial guidance system, corrected by stellar information provided by the ACN, guided the missile into the target.

launch two SNARKs
The SNARK B-62 Operations Section assisted with the conditioning and installation of the rocket boosters used on both flights, and it took some satisfaction in knowing that the boosters performed well during both launches.

Model N-69C missiles

Though the "C" model was used for terminal dive testing, it could fly for about two hours before reaching its "dump" point; this gave Northrop a chance to pick up some flight data missed on earlier N-69A and N-69B flights. The first "C" model launched from the Cape was also the first SNARK to be equipped with new solid rockets rated at 130,000 pounds of thrust.

SNARK guidance test flights
A few of the missiles failed to accomplish all of their objectives, and one N-69D had the dubious distinction of flying completely off the Range "without permission." Following its launch on 5 December 1956, the delinquent SNARK failed to respond to every external guidance command sent to it. After disregarding all destruct commands sent to it, the missile finally crash-landed harmlessly in the jungles of Brazil. Apparently, the missile's destruct system had been rendered inoperative due to a power failure; the destruct system on later SNARK missiles was modified to avoid this type of incident, and no other missiles went AWOL (Away Without Official Leave) in later years.

specialized training
A few of the 6555th's officers and airmen had been integrated into the Northrop Field Test Crew, but most of the 80 military personnel assigned to SNARK activities had been relegated to support roles up to that point in the program.

Air Force decided
This decision was based on the likelihood that intercontinental ballistic missiles would render the SNARK obsolete by the early 1960s. Previously, the 556th crew training program was to be completed by June 1960. Subsequently, training had to be completed by the end of December 1959.

crew training
Under an informal agreement between Air Training Command and AFMTC, one officer and five airmen were sent to AFMTC in March 1959 and attached to the 6555th Guided Missiles Squadron to train officers and airmen for SAC's SNARK unit at Presque Isle, Maine. The first graduating class consisted of 8 officers and 72 airmen.

XSM-64 test vehicle
When requirements for the NAVAHO were firmed up in the early 1950s, the XSM-64 missile was expected to weigh about 65,000 pounds and its booster was expected to weigh about 71,700 pounds. Some evidence suggests that the missile and booster eventually grew to 70,000 pounds and 90,000 pounds respectively, but the two 120,000-pound thrust rocket engines were powerful enough to boost the combined weight of the NAVAHO in either case.

landing strip
This $2,000,000 landing field became known as the "Skid Strip." In later years, the Skid Strip was widened to 300 feet, resurfaced and expanded to include a taxiway and a parking apron for transports arriving with missile and spacecraft components. Eventually, the landing area accommodated the heaviest cargo carriers in the Air Force. A small control tower and a modest fire and crash rescue capability complemented the airfield.

The 6555th

Chapter Two Endnotes

1. AFMTC History, 1 July - 31 December 1953, p. 262; AFMTC History, 1 July - 31 December 1959, p. 149.

2. AFMTC History, 1 January - 30 June 1953, pp. 131, 132.

3. General Order Number 8, HQ ARDC, 14 May 1951; General Order Number 24, HQ LRPGD, 14 May 1951; 6555th Guided Missile Wing History, June 1951, p. 7; 6555th Guided Missile Wing History, September - October 1951, p. 29; AFMTC History, 1 January - 30 June 1952, pp. 43, 211, 212, 215, 218-222, 271, 276, 280, 281; AFMTC History 1 July - 31 December 1952, p. 36; General Order Number 21, HQ ARDC, 17 February 1953; General Order Number 5, HQ AFMTC, 26 February 1953; AFMTC History, 1 January - 30 June 1953, pp. 33, 34.

4. 6555th Guided Missile Wing History, June 1951, pp. 3, 12, 22; Regulation 24-1, AFMTC, "6555th Guided Missile Wing," 29 August 1951; 6555th Guided Missile Wing History, January - February 1952, pp. 23, 26, 32; 6555th Guided Missile Wing History, November - December 1951, pp. 18, 25, 37; 6555th Guided Missile Wing History, May - June 1952, pp. 2, 16, 27, 30, 33, 37, 43.

5. AFMTC History, 1 January - 30 June 1952, pp. 170, 171; 6555th Guided Missile Wing History, pp. 5, 6.

6. AFMTC History, 1 January - 30 June 1952, p. 422; 6555th Guided Missile Wing History, January - February 1952, pp. 22, 27.

7. 6555th Guided Missile Wing History, November - December 1951, pp. 36, 37; 6555th Guided Missile Wing History, May - June 1952, pp. 39, 45; AFMTC History, 1 January - 30 June 1952, p. 429; AFMTC History, 1 July - 31 December 1952, p. 439; 6555th Guided Missile Wing History, March - April 1952, pp. 29-31.

8. TAC Fact Sheet, "MATADOR," undated; Office of Information Services, Patrick AFB, "Fact Sheet," 3 March 1955; AFMTC History, 1 January - 30 June 1952, pp. 176-178.

9. AFMTC History, 1 January - 30 June 1952, pp. 173, 174.

10. Ibid., p. 175; AFMTC History, 1 January - 30 June 1953, pp. 148, 149.

The 6555th Chapter II Endnotes

11. 6555th Guided Missile Wing History, September - October 1951, pp. 23, 41; AFMTC History, 1 January - 30 June 1952, pp. 172, 174, 175; 6555th Guided Missile Wing History, March - April 1952, p. 28; 6555th Guided Missile Wing History, May - June 1952, p. 27.

12. AFMTC History, 1 January - 30 June 1952, pp. 175, 197, 198, 205; AFMTC History 1 January - 30 June 1953, pp. 33, 34.

13. AFMTC History, 1 January - 30 June 1953, pp. 33, 34, 125, 126, 332; AFMTC History, 1 July - 31 December 1953, pp. 34, 209, 210.

14. AFMTC History, 1 July - 31 December 1953, pp. 212, 213, 233, 281; AFMTC History, 1 July - 31 December 1954, p. 199.

15. AFMTC History, 1 July - 31 December 1953, pp. 283, 284, 286, 287.

16. AFMTC History, 1 January - 30 June 1954, pp. 28, 29, 186-188; AFMTC History, 1 July - 31 December 1954, p. 31; Interview, Mr. Robert F. Friedmann (former motor transport officer with the 69th PBS), with Mark C. Cleary, 17 May 1991.

17. AFMTC History, 1 July - 31 December 1954, pp. 27, 28, 31, 39, 200.

18. AFMTC History, 1 January - 30 June 1955, pp. 187, 188, 214, 236, 356; AFMTC History, 1 July - 31 December 1955, pp. 136, 139-141.

19. Whipple, Marven R., AFETR History Office, "List of 6555th Commanders," o/a 1 January 1967; AFMTC History, 1 July - 31 December 1959, pp. 19, 25; General Order Number 79, HQ ARDC, 4 August 1959; General Order Number 99, HQ ARDC, 17 September 1959; General Order Number 238, HQ ARDC, 14 December 1959.

20. AFMTC History, 1 January - 30 June 1956, pp. 151-161, 163, 164; AFMTC History, 1 July - 31 December 1956, pp. 151-154, 156-158, 231; AFMTC History, 1 January - 30 June 1957, pp. 153, 154.

21. AFMTC History, 1 January - 31 December 1957, p. 155; AFMTC History, 1 July - 30 June 1957, pp. 157, 159; AFMTC History, 1 January - 30 June 1958, p. 136; AFMTC History, 1 July - 31 December 1958, pp. 151, 152; AFMTC History, 1 January - 30 June 1959, pp. 146, 147; AFMTC History, 1 July - 31 December 1959, pp. 149, 150; Crespino, Janice E., ESMC/HO, "Launches From The Eastern Test Range, 1950 - 1990," April 1991, p. 20.

22. AFMTC History, 1 January - 30 June 1952, pp. 175, 176; AFMTC History, 1 July - 31 December 1959, pp. 151-153.

23. History of the MACE Operations Division, Directorate of Operations, 6555th Test Wing (Development), 21 December 1959 - 31 March 1960, p. 1.

24. Ibid., p. 2; History of the MACE Operations Division, Directorate of Operations, 6555th Test Wing (Development), 1 April - 30 June 1960, p. 1; History of the MACE Operations Division, Directorate of Operations, 6555th Test Wing (Development), 1 July - 31 December 1960, pp. 1, 2; History of the MACE Operations Division, Directorate of Operations, 6555th Test Wing (Development), 31 December 1960 - 1 July 1961, p. 1; Crespino, "Launches," pp. 14, 15; History of the MACE Weapons Branch, Deputy for Ballistic Systems, 6555th ASTW, 1 Jan - 30 Jun 62, p. 1

25. LRPGD History, 1 July - 31 December 1950, pp. 156, 157; LRPGD History, 1 January - 30 June 1951, p. 130; 6555th Guided Missile Wing History, June 1951, pp. 6, 7; 6555th Guided Missile Wing History, January - February 1952, pp. 24, 25; AFMTC History, 1 January - 30 June 1952, pp. 231-235; AFMTC History, 1 July - 31 December 1953, p. 269; Crespino, "Launches," pp. 13, 14.

26. Crespino, "Launches," p. 7; AFMTC History, 1 January - 30 June 1952, pp. 262, 266, 279; AFMTC History, 1 January - 30 June 1953, pp. 228, 229.

27. AFMTC History, 1 January - 30 June 1953, pp. 235, 236; AFMTC History, 1 July - 31 December 1953, pp. 243, 244, 291; AFMTC History, 1 July - 31 December 1956, pp. 42, 168, 169, 172, 173; AFMTC History, 1 July - 31 December 1958, p. 16, AFMTC History, 1 January - 30 June 1959, p. 156.

28. AFMTC History, 1 July - 31 December 1958, pp. 26, 160, Crespino, "Launches," pp. 7, 8.

29. AFMTC History, 1 July - 31 December 1951, p. 160; AFMTC History, 1 January - 30 June 1952, pp. 244, 245, 247, 252; 6555th Guided Missile Wing History, August 1951, pp. 4-7; 6555th Guided Missile Wing History, July 1951, p. 13.

30. AFMTC History, 1 January - 30 June 1952, pp. 43, 246, 252, 253, 258; AFMTC History, 1 July - 31 December 1951, p. 152; 6555th Guided Missile Wing History, May - June 1952, p. 34.

31. AFMTC History, 1 January - 30 June 1953, pp. 206, 208.

32. AFMTC History, 1 July - 31 December 1952, pp, 251-253; Crespino, "Launches," p. 47; AFMTC History, 1 January - 30 June 1953, pp. 207, 214, 215; AFMTC Report to Management, September 1953, p.3.

33. AFMTC History, 1 January - 30 June 1952, p. 247; AFMTC History, 1 July - 31 December 1953, pp. 255, 256, 257-260; AFMTC History, 1 January - 30 June 1954, pp. 150-152; AFMTC History, 1 January - 30 June 1957 p. 158.

34. AFMTC History, 1 January - 30 June 1955, pp. 264, 265, 359; AFMTC History, 1 July - 31 December 1955, pp. 154, 155; AFMTC History, 1 January - 30 June 1956, p. 215; AFMTC History, 1 July - 31 December 1956, pp. 159, 160, 163, 164, 167.

35. AFMTC History, 1 July - 31 December 1956, pp. 159, 160, 163, 164, 167; AFMTC History, 1 January - 30 June 1957, pp. 161-164; AFMTC History, 1 July - 31 December 1957, pp. 161, 162, 164, 165; AFMTC History, 1 January - 30 June 1958, pp. 137, 138, 140, 141.

36. AFMTC History, 1 January - 30 June 1957, pp. 167, 169.

37. AFMTC History, 1 July - 31 December 1957, pp. 29, 30, 163, 164; AFMTC History, 1 January - 30 June 1958, p. 138, 140, 141; AFMTC History, 1 July - 31 December 1958, p. 154.

38. AFMTC History, 1 July - 31 December 1958, pp. 154, 157; AFMTC History, 1 January - 30 June 1959, pp. 30, 31, 149, 150; Del Papa, E. Michael, (et al), From Snark to Peacekeeper, A Pictorial History of Strategic Air Command Missiles, SAC History Office, 1 May 1990, pp. 2, 5, 77; AFMTC History, 1 July - 31 December 1959, p. 160.

39. AFMTC History, 1 July - 31 December 1959, pp. 155, 158, 159; Crespino, "Launches," p. 49.

40. Del Papa, Snark to Peacekeeper, pp. 5, 77.

41. AFMTC History, 1 July - 31 December 1957, pp. 166, 167; AFMTC History, 1 January - 30 June 1958, p. 144; AFMTC History, 1 July - 31 December 1958, pp. 161, 162.

42. AFMTC History, 1 January - 30 June 1952, pp. 285, 286; AFMTC History, 1 January - 30 June 1953, p. 130; ESMC History, 1 October 1989 - 30 September 1990, p. 93; AFMTC History, 1 July - 31 December 1953, pp. 278, 322, 323.

43. AFMTC History, 1 January - 30 June 1956, p. 55; AFMTC History, 1 January - 30 June 1955, p. 410; AFMTC History, 1 July - 31 December 1955, p. 259, 322; AFMTC History, 1 July - 31 December 1954, p. 283; Crespino, "Launches," p. 22.

44. AFMTC History, 1 July - 31 December 1956, pp. 176, 182, 183; AFMTC History, 1 January - 30 June 1957, pp. 173-175; Crespino, "Launches," p. 22.

45. AFMTC History, 1 January - 30 June 1957, pp. 175, 176, 181; AFMTC History, 1 July - 31 December 1957, pp. 166-168; AFMTC History, 1 January - 30 June 1958, p. 143.

46 and 47. AFMTC History, 1 January - 30 June 1958, p. 144; AFMTC History, 1 July - 31

Dec 1958, pp 161-164; AFMTC History, 1 Jan - 30 Jun 59, p 158; Crespino, "Launches," pp 22, 23

The 6555th, Chapter III, Section 1

The 6555th's Role in the Development of Ballistic Missiles

Ballistic Missile Test Organizations and Commanders

As MATADOR flight testing got underway at Cape Canaveral in the summer of 1951, Air Force planners redoubled their efforts to develop the ballistic missile as a logical successor to the pilotless bomber. Convair was awarded an Air Force contract to study the merits of the ballistic missiles in relation to aerodynamic missiles, and, in September 1951, Convair proposed a ballistic missile along the lines suggested by Consolidated-Vultee's rocket experiments in the late 1940s (i.e., a lightweight pressurized booster with swiveling engines for directional control and a separable nose cone to simplify atmospheric reentry problems). An ad hoc committee of the Air Force's Scientific Advisory Board supported Convair's proposal on the grounds that it was technically feasible, and Convair presented the Air Force with a plan in 1953 for rapid development of the missile.[1]

In October 1953, an 11-member Air Force panel of experts was formed under Dr. John von Neumann to evaluate strategic missile programs. This Strategic Missiles Evaluation Committee (SMEC) was nicknamed the "Teapot Committee" in a light-hearted gesture that belied the seriousness of its work. In February 1954, the Teapot Committee recommended a "radical reorganization" of America's ballistic missile effort to catch up with the Soviet Union in long-range ballistic missile development: the Soviets were clearly ahead of the Americans in heavy ballistic missiles by this time, and they had tested their first H-Bomb successfully in August 1953. The Committee noted that a recent breakthrough in nuclear warhead design offered the U.S. a shortcut, making a relatively lightweight (240,000-pound) intercontinental ballistic missile (ICBM) possible within eight years. This missile would only weigh half as much as the ATLAS ICBM proposed by Convair (e.g., 450,000 pounds), and the H-Bomb's extra "punch" would allow designers to loosen proposed target accuracy requirements from 1,500 feet to approximately three miles. (Accuracy requirements were loosened further -- to five miles -- after the Atomic Energy Commission predicted it could develop a one-megaton warhead light enough to be carried on the 240,000-pound version of the ATLAS.) Given those parameters, the ATLAS' five-engine configuration could be trimmed down to a three-engine, booster-sustainer (1 and 1/2 stage) design.[2]

Based on the Teapot Committee's recommendations, RAND studies and successful lightweight H-Bomb tests in 1953 and 1954, the Air Force Vice Chief of Staff (General Thomas D. White) assigned the Air Force's highest research and development priority to the ATLAS project (Weapon System 107A-1) on 14 May 1954. On July 1st, the Air Research and Development Command established the Western Development Division (WDD) under the command of Brigadier General Bernard A. Schriever to manage the ATLAS project. Toward the end of August 1954, General Schriever recommended that the

Ramo-Wooldridge Corporation be given responsibility for technical direction and systems engineering for ATLAS, and Ramo-Wooldridge became an indispensable partner in the WDD's supervision of contracts for the ATLAS and later ballistic missile programs. A full "go-ahead" for the ATLAS design was ordered in January 1955, and the TITAN (Weapon System 107A-2) was added to the ICBM effort to trail behind the ATLAS' development program by about a year as a "hedge against failure."[3]

MAJOR GENERAL BERNARD A. SCHRIEVER
As WDD Commander In 1956

To counter the likelihood that the Soviets would have ballistic missiles before the U.S. could field the ATLAS or TITAN ICBMs, the Air Force awarded a research and development contract to the Douglas Aircraft Company on 27 December 1955 for the THOR intermediate-range ballistic missile (IRBM). This effort was designed to get a strategic ballistic missile (Weapon System 315A) into the West's inventory as soon as possible. Like the ATLAS and TITAN, THOR requirements were "frozen" early in the development process to avoid further delays, and the various missile components and support equipment for each weapon system were developed concurrently to insure the earliest initial operating capability (IOC) for each type of missile. Bureaucratic red tape and funding delays were also reduced significantly after the Assistant Secretary of the Air Force for Research and Development (Trevor Gardner) had the Deputy for Budget and Program Management (Hyde Gillette) set up a committee to streamline administration. The new "Gillette Procedures" were approved in November 1955, and they cut the number of official agency review levels for ballistic missiles from 42 to 10. The Western Development Division became the Air Force Ballistic Missile Division (AFBMD) on 1 June 1957, and it continued to manage a whole family of ballistic missile programs, reconnaissance satellite projects and at least one solid propellant rocket project.[4]

Because most ballistic missile components were being tested elsewhere in the United States, the Western Development Division only required a very small liaison office at AFMTC from the middle of August 1955 through the end of April 1956. By the middle of 1956, ballistic missile flight tests were anticipated at Cape Canaveral in 1957, and the liaison office was replaced by the Western Development Division Field Office on 1 May 1956. (Ramo-Wooldridge activated its own Flight Test Office at AFMTC on the same date to provide technical assistance.) Though the Field Office only had three officers and four civilians assigned to its operations when it opened for business in May, it grew slowly and steadily to 49 officers, eight airmen and 21 civilians by December 1959. The ATLAS, TITAN and THOR field testing programs were assigned to the Field Office initially, and the X-17 Research Test Vehicle and the MIDAS satellite project were added shortly thereafter. The Field Office's X-17 Branch was phased out on 1 May 1957, but its personnel were given other duties, including the MINUTEMAN (Weapon System 133A) field test program, which required the creation of a MINUTEMAN project division in January 1959. (On 1 December 1957, the Field Office was renamed the Air Force Ballistic Missile

Division's Office of the "Assistant Commander for Missile Tests," but it continued to function as a field office.) Many of the Air Force Ballistic Missile Division's contractors and sub-contractors maintained their own field offices at AFMTC, and it was the Field Office's function to provide a working liaison between General Schriever's Division and AFMTC.[5]

Since the Assistant Commander for Missile Tests' resources were reassigned to the 6555th Guided Missile Group (Test and Evaluation) when the latter was redesignated the 6555th Test Wing (Development) on 21 December 1959,* we should make some mention of the Field Office's commanders as well as their successors in the 1960s. Lieutenant Colonel Charles G. Mathison became the Chief of the WDD Liaison Office in August 1955, and he continued to serve as General Schriever's Assistant Commander for Missile Tests in the Western Development Division's Field Office through 8 July 1956. Lieutenant Colonel Mathison was succeeded by Colonel Henry H. Eichel on July 9th, and Colonel Eichel continued in that capacity after the Field Office was renamed the Office of the Assistant Commander for Missile Tests in December 1957. Colonel Eichel also became the first commander of the 6555th Test Wing (Development) on 21 December 1959, when the 6555th Guided Missile Group (Test and Evaluation) was removed from AFMTC, redesignated and assigned to the Air Force Ballistic Missile Division. Colonel Paul R. Wignall succeeded Colonel Eichel as the 6555th's Commander on 13 June 1960, and he commanded the Wing for the next two and one-half years. Colonel Harold G. Russell commanded the 6555th from 1 December 1962 through 2 August 1964. He was succeeded by Colonel Otto C. Ledford, who served through 14 September 1967. Colonel Marc M. Ducote assumed command briefly in September, but he was succeeded by Colonel Herbert J. Holdsambeck on September 26th. Colonel Holdsambeck continued to command the 6555th until 9 August 1969, whereupon Lieutenant Colonel Robert H. Reynolds assumed command temporarily until Colonel Davis P. Parrish's arrival on 24 September 1969. Colonel Parrish continued to command the 6555th through 23 August 1972.[6]

COLONEL PAUL R. WIGNALL

COLONEL HAROLD G. RUSSELL

COLONEL OTTO C. LEDFORD

COLONEL HERBERT S. HOLDSAMBECK

COLONEL DAVIS P. PARRISH

The 6555th's manpower and mission in the last half of the 1950s should also be mentioned. When the Western Development Division's liaison office opened in August 1955, the 6555th Guided Missile Squadron had 11 officers and 135 airmen assigned to various aerodynamic missile programs. During this period, the 6555th: 1) launched MATADORS, 2) supported the BOMARC, and 3) prepared for the day when blue suit crews would begin launching SNARKs. While the Liaison Office's presence grew, the 6555th's strength remained relatively stable (except for the last half of 1958) at about a dozen officers and 145 airmen. After the 6555th was redesignated a Wing and assigned to the Air Force Ballistic Missile Division on 21 December 1959, it picked up technical management for the Air Force's ballistic missile flight test programs at the Cape, and its mission was expanded to include the attainment of a military launch, test and evaluation capability for ballistic missiles. The Wing gained the resources of the Assistant Commander for Missile Tests, and it had a force of 71 officers, 159 airmen and 21 civilians by the end of 1959.[7]

The 6555th, Chapter III, Section 2

The 6555th's Role in the Development of Ballistic Missiles

The Eastern Test Range in the 1950s

The Eastern Test Range also changed to meet new ballistic missile program requirements. Though the SNARK and NAVAHO prompted expansion of the Eastern Test Range to Ascension Island in the mid-1950s, those winged missiles were not destined to become the principal users of the Range's most distant outposts. After what appeared to be a slow start, ballistic missile programs took root at the Cape and quickly dominated the Range after 1957. At the beginning of 1956, the Eastern Test Range extended from Cape Canaveral to the instrumentation station at Mayaguez, Puerto Rico -- a distance of approximately 1,000 miles. In addition to facilities at the Cape and Mayaguez, range stations were located near Jupiter, Florida, on the islands of Grand Bahama, Eleuthera, San Salvador, Mayaguana and Grand Turk. Instrumentation consisted of telemetry receiving stations, radar tracking sites, optical systems, command/destruct equipment, timing systems, communications stations and various types of recording equipment. Ascension Island had just been selected as the terminal point for a 5,000 nautical mile range, and St. Lucia and Fernando de Noronha were selected as intermediate stepping stones in the instrumentation chain a little later on. Picket ships were also required to fill in the gaps between St. Lucia, Fernando de Noronha and Ascension. While operations involving the MATADOR, BOMARC, SNARK, NAVAHO, the X-17 and the Army's JUPITER were underway at the Cape in 1956, preparations continued for the missile programs that would dominate the Range toward the end of the 1950s -- the Air Force's THOR, ATLAS, and TITAN, and the Navy's POLARIS.[8]

MAIN BASE, GRAND BAHAMA ISLAND

SAN SALVADOR - 1954

*CENTRAL CONTROL BUILDING
SAN SALVADOR*

MAIN BASE, ASCENSION - April 1959

FIVE RANGE INSTRUMENTATION SHIPS IN WET STORAGE AT TRINIDAD ISLAND - 1958

SIDE VIEW OF THE RANGE SHIP ROSE KNOT - 1958

RACKS OF TELEMETRY EQUIPMENT ABOARD A RANGE SHIP - 1958

In general, all missile test programs had certain requirements in common. Sensors in the launch area were arrayed to gather data on high angle and low angle launches during the developmental phase of each missile program. That instrumentation measured the missile's position, velocity, acceleration, altitude and attitude to verify stability and control characteristics as the missile lifted off the pad. Most missile test programs could use the same instrumentation, though it might have to be rearranged or reconfigured to meet specific test requirements (e.g., picket ships and optical sensors). However, unlike aerodynamic "cruise" missiles, ballistic missiles had critical staging sequences when rocket engines shut down and booster segments dropped off. During those portions of the flight, a high degree of tracking accuracy was required for ballistic missiles. Near the end of a flight, data requirements tended toward higher accuracies for winged and ballistic missiles alike: winged missiles encountered stability problems during their terminal dives, and ballistic missile reentry vehicles coped with the stresses of reentering the atmosphere. In both instances, contractors needed very precise information on those events. On the whole, ballistic missile programs required a more sophisticated range, but many sensors procured for aerodynamic missile tests also served ballistic missile programs in later years.[9]

The Range was equipped with single-point radars initially. Those radars were called "MOD I" radars, because they were derived from the old SCR-584 radar system. The MOD I was the most economical solution to aerodynamic missile test requirements, but the AZUSA continuous wave tracking system was introduced in the mid-1950s to meet more stringent ballistic missile test requirements. MOD I and AZUSA radars had a distinct advantage over Doppler Velocity and Position (DOVAP) radars in that they required a minimum number of radar sites and operating crews. The increased emphasis on ballistic missiles and their higher accuracy requirements forced the Range to upgrade its radars around the middle of the 1950s, and the MOD I radars were replaced with MOD II S-Band radars supplied by the

Reeves Instrument Company. Those, in turn, were replaced by FPS-16, radars at Patrick, Cape Canaveral, Grand Bahama, San Salvador and Ascension in late 1950s and the early 1960s. Larger C-Band radars were added at Patrick, Grand Bahama, Grand Turk, Ascension and Merritt Island to support APOLLO and MINUTEMAN launches later on. Modifications to individual radars continued through the 1980s.[10]

*RADAR DISH ABOARD RANGE
SHIP
December 1958*

By the end of 1957, the Range's optical systems included long-range tracking telescopes, cinetheodolite systems, infrared tracking equipment and ribbon-framed cameras in 16 mm, 35 mm and 70 mm formats. CZR-1 ribbon-framed cameras covered the missile during the first 1,000 feet of a launch, and cinetheodolites followed the flight out to about 20 miles. Wild BC-4 1958 ballistic cameras captured optical data beyond the tracking radars' beamwidths, and they also obtained time-position data for ballistic missile staging events and reentry phenomena. Long-range tracking telescopes provided coverage as far as 200 miles downrange, depending on the weather, air turbulence and the time of day. Infrared cameras tracked missiles in the dark.[11]

More than 60 percent of the test data obtained on a missile flight in the late 1950s was gathered by two types of telemetry systems: 1) the frequency modulation system (FM) and 2) the pulse duration modulation system (PDM/FM). Both radio-based systems provided information on the internal characteristics and performance of a missile in flight. While the PDM/FM transmitter was smaller and lighter than the FM transmitter (making it ideal for small missile applications), the FM system had better channel frequency response. There was a telemetry station on Grand Turk in 1956, and other telemetry sites were activated when stations were opened at Mayaguez, Antigua and Ascension in 1956 and 1957. Radio communications, a timing system and a submarine cable system also tied the Range's stations together to insure coordinated coverage of each missile's flight. The Range was operated and maintained for the Air Force by Pan American World Services and RCA from 1954 through most of the 1980s, and by Computer Sciences Raytheon and Pan Am from October 1988 onward.[12]

AZUSA ANTENNA FIELD - 1954

INFLATED ENCLOSURES FOR AZUSA ANTENNAS - 1954

MOD II RADAR DISH WITH CAMERA ATOP
CENTRAL CONTROL BUILDING
Cape Canaveral 1956

MOD II RADAR REMOTE CONTROL UNIT

MOD II RADAR EQUIPMENT VANS - 1956

MOD II RADAR CONSOLES - 1956

FPS-16 RADAR AT CAPE CANAVERAL - 1963

CZR-1 CAMERA AND MOUNT - 1960

80-INCH ZOOMAR ON MK45 MOUNT WITH 35 MM MITCHELL CAMERA - 1960

BC-4 BALLISTIC CAMERA - 1961

RECORDING OPTICAL TRACKING INSTRUMENT (ROTI) - 1958

INFRARED TRACKER MOUNTED ON FPS-16 RADAR - 1961

TLM-18 TELEMETRY TRACKER - 1961

60-FOOT TRACKING ANTENNA

The 6555th, Chapter III, Section 3

The 6555th's Role in the Development of Ballistic Missiles

Ballistic Missile Test Objectives

POLARIS OPERATIONS AT PAD 3,
Cape Canaveral, April 1957

Ballistic missile processing was also a contractor-oriented function, but it became a major part of the 6555th's mission at the end of 1959. At that time, component testing took place in laboratories, factories and test centers in various parts of the United States. Even if an item was considered highly reliable, it received an exhaustive series of new tests to prove its mettle before it was added to a new missile's inventory of parts. Once a missile arrived at Cape Canaveral, it received a meticulous inspection and a thorough series of pre-flight tests. For almost all missiles, this included on-the-ground engine checks and static firings before launch day. Ballistic missile flight test programs advanced in much the same manner as aerodynamic missiles had in the early 1950s. On early flights, the airframe, propulsion system and autopilot were measured against established standards for structural integrity and responsiveness. After the basic elements of the missile were tested successfully, more complex items such as the guidance system, reentry vehicles, simulated warheads and fuzes were evaluated until the complete weapon system proved itself. Specific primary objectives were established for each missile flight test, and the degree of success or failure was judged by the extent to which data relative to those primary objectives were obtained. Thus, an apparent "failure" might constitute a very successful test -- depending on how well the flight met the intended primary objectives. Moreover, it was usually possible to establish the exact cause of a genuine flight failure by analyzing the data collected by range instrumentation. On the basis of this information, remedial action (including major changes at the contractor's plant) could be accomplished before the next launch. The thoroughness of the experimental test program had a direct bearing on the quality of the production line missile, and, at times, missiles were subjected to conditions beyond their designed tolerances to determine the maximum stress they could tolerate before structural failure. After its reassignment to the Air Force Ballistic Missile Division in 1959, the 6555th became keenly interested in all those aspects of Air Force ballistic missile testing at Cape Canaveral.[13]

The 6555th, Chapter III, Section 4

The 6555th's Role in the Development of Ballistic Missiles

The Thor Ballistic Missile Test Program

The Air Force Missile Test Center became involved with the THOR (Weapon System 315A) program in the fall of 1954, after ARDC ordered development of that missile "as soon as possible." The Wright Air Development Center sponsored the missile initially, and AFMTC hosted the THOR at Cape Canaveral. Following a series of meetings between AFMTC and Western Development Division officials in February and March 1955, support requirements were worked out for two launch pads, a blockhouse, a guidance site, one service stand, airborne guidance test equipment, housing and messing facilities. The THOR was given equal priority with the ATLAS in December 1955, and the Western Development Division became the missile's new sponsor at that time. The first THOR test missile was launched in January 1957, and, by that time, THOR launching facilities consisted of a blockhouse, one launching pad with a service tower (i.e., Pad 17B), a second partially finished launch pad (17A) and a 40,000-square-foot assembly building (Hangar M). The principal contractors were Douglas Aircraft Company (for the airframe), Bell Telephone Laboratories (for the radio-inertial guidance system), A.C. Spark Plug (for the more advanced all-inertial guidance system), General Electric (for the nose cone) and North American Aviation (for the rocket motors). The THOR weighed 110,400 pounds, and it was 62.5 feet long and 8 feet in diameter. It was propelled by a single rocket motor rated between 135,000 pounds and 150,000 pounds of thrust.[14]

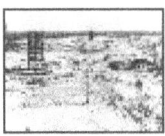

COMPLEX 17 UNDER CONSTRUCTION - 1956

MISSILE ASSEMBLY BUILDING M (LEFT) AND STEEL STRUCTURE OF HANGAR L (RIGHT) - 1956

The first THOR was launched from Cape Canaveral on 25 January 1957. Engine ignition and main stage operation were normal at launch, but a liquid oxygen valve failed almost immediately after lift-off, and the missile slipped back through Pad 17B's launcher ring to explode on the deflector plate below. A low order explosion and fire destroyed the missile and damaged the pad, and the launch failure delayed the next launch until mid-April 1957. The second THOR launch, on April 19th, was more successful, but a third missile (Number 103) exploded on the pad on May 21st after its fuel tank ruptured five minutes before its intended launch. Once again, Pad 17B had to be refurbished, but Pad 17A was completed in July 1957, so 17A was used for the fourth THOR on August 30th. (That launch was successful, but the missile broke in half 93 seconds into the flight and plunged into the Atlantic about 20 miles from the launch site.) The fifth THOR was launched from Pad 17B on 20 September 1957, and it met all its test objectives. Unfortunately, Pad 17A received its own baptism in fire when the sixth THOR lost thrust on October 3rd, fell back through the launching ring and burned. Two other launches on October 11th and October 24th were successful, and they concluded the THOR's airframe/propulsion testing phase at the Cape. Guidance system tests began with a partially successful THOR flight from Pad 17B on 7 December 1957. A subsequent flight from Pad 17A tested the THOR's all-inertial guidance system on December 19th. It was completely successful.[15]

THOR NUMBER 101 ON LAUNCH STAND BEFORE LAUNCH

THOR 101 LAUNCH FAILURE
25 January 1957

THOR 101 EXPLOSION
25 January 1957

REMOVING THOR 101 WRECKAGE

THOR NUMBER 103 IN SERVICE TOWER, PAD 17B
6 May 1957

DAMAGE TO PAD 17B FOLLOWING EXPLOSION OF THOR 103
21 May 1957

THOR nose cone separation tests began with a flight from Pad 17B on 28 February 1958, but the second launch in that series ended in another explosion and fire on Pad 17B on April 19th. The third nose cone test was flown successfully on June 13th, and a THOR tactical launcher performed well during a guidance system test launch offsite (e.g., Pad 18B) on 4 June 1958. Test flights to validate overall refinements in the THOR got underway on 5 November 1958, and five of those flights were launched from Pad 17B or 18B by the end of the year. Three of the five met their test objectives, and so did eight of the next nine THORs launched from 17B or 18B between 30 January and the end of June 1959. An 82-foot-long variant of the THOR, known as the THOR-ABLE, was also used in a dozen flights from Pad 17A between 23 April 1958 and 12 June 1959 to test two different ablative nose cone designs for the ATLAS program. Eleven out of 14 THORs met their flight test objectives in the last half of 1959.[16]

THOR-ABLE LAUNCH FROM PAD 17A
23 January 1959

Though contractors launched many THORs at Cape Canaveral, the Air Force conducted its ballistic missile Combat Training Launch (CTL) operations at Vandenberg Air Force Base. The first missile launched from Vandenberg was also the first THOR CTL operation, and it was completed successfully by a blue suit (SAC) missile crew on 16 December 1958. The British participated in THOR launches at Vandenberg and Cape Canaveral a little later on, and they increased their presence at Vandenberg and the Cape as blue suit and contractor support steadily diminished over the next two and one-half years. As testing continued at the Cape, THORs began to arrive in the United Kingdom in September 1958, and 60 THOR launch sites (assigned to four THOR squadrons) went on alert between June 1959 and April 1960. THOR launch operations were performed exclusively by Royal Air Force personnel after June 1961. Since the THOR's mission could be assumed by other weapon systems after 1962, the THORs were pulled out of Great Britain between November 1962 and August 1963, and they were returned to the United States. The THOR was matched with several different high energy upper stages in the late 1950s, and, as of this writing, the THOR booster continues to serve as the first stage of a space vehicle known as the DELTA II used for Global Positioning Satellite (GPS) and commercial space launch operations.[17]

The 6555th, Chapter III, Section 5

The 6555th's Role in the Development of Ballistic Missiles

The Atlas Ballistic Missile Program

The U.S. also made great strides with ICBM programs in the late 1950s and early 1960s, beginning with the ATLAS. The ATLAS' development was a much larger enterprise than the THOR program, but its flight test program moved ahead quickly once the missile arrived at the Cape. Like THOR, ATLAS involved several major contractors: Convair (General Dynamics) was responsible for the ATLAS' airframe; North American had the contract for the missile's rocket engines; General Electric had the contract for the nose cone, and it shared the missile's guidance system contract with the A.C. Spark Plug Company. In a related effort, the Lockheed Aircraft Corporation conducted ATLAS reentry vehicle research under the X-17 program at Cape Canaveral between May 1955 and the end of March 1957. Following completion of the X-17 flight test program in March, Convair proceeded with the ATLAS development program, which was scheduled to advance through four series of flight tests:[18]

- Series A - Airframe and propulsion tests, employing seven 181,000-pound test missiles between June and the end of December 1957.
- Series B - Booster separation and propulsion tests, employing three 248,000-pound test missiles between January and the end of March 1958.
- Series C - Guidance and nose cone tests, employing eighteen 243,000-pound test missiles between April and the end of November 1958.
- Series D - Operational tests of ATLAS prototype (i.e., the complete missile), employing twenty-four 243,000-pound ATLAS prototypes between December 1958 and the end of July 1959.

X-17 ON TRANSPORT TRAILER 1957

LOWERING X-17 TRAILER August 1957

CONSTRUCTION PROGRESS ON MISSILE ASSEMBLY BUILDING K

1956

MISSILE ASSEMBLY BUILDING K

Late 1956

BLOCKHOUSE 14 UNDER CONSTRUCTION
May 1956

BLOCKHOUSE 14 FOLLOWING CONCRETE POUR

July 1956

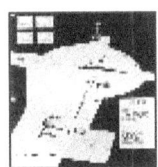
TYPICAL LAUNCH SITE LAYOUT
Cape Canaveral 1958

The missiles used in all four series were 75 feet long and 10 feet in diameter. The lightest (Series A) missiles were only expected to fly 460 nautical miles downrange, but each succeeding series would be flown further (e.g., 3,000 nautical miles for Series B, 4,500 nautical miles for Series C and 5,500 nautical miles for Series D). Though the overall flight schedule was dependent on the timely completion of ATLAS facilities at the Cape, the first two ATLAS launch pads, a missile storage building (Building "K"), a missile guidance facility and a data collection equipment station were completed by the end of 1956. The first Series A test missile was launched from Pad 14 on 11 June 1957, but the ATLAS' rocket engines lost thrust approximately 24 seconds into the flight. The missile performed a "couple of loops" and fell through its own trail of fire before a Range Safety Officer sent a command to destroy it less than a minute after lift-off. Despite the missile's failure, the flight met some of the test objectives for the airframe and the launch system.[19]

ERECTION OF AN ATLAS MISSILE AT PAD 14

1957

Two more Series A missiles were launched from Pad 14 on 25 September and 17 December 1957.

During the first of those flights, the propulsion system functioned normally for approximately 32 seconds before a liquid oxygen regulator problem reduced thrust and finally cut all power. The third ATLAS flight on December 17th was completely successful, and another successful flight marked Pad 12's debut as an ATLAS launch site on 10 January 1958. Pad 14 supported two other Series A launches on 7 February and 5 April 1958, and two more Series A missiles were launched from Pad 12 on 20 February and 3 June 1958. The mission on June 3rd was the final flight in the ATLAS "A" Series.[20]

The first Series B ATLAS was launched from a third site --Pad 11 --on 19 July 1958. The missile lost thrust 43 seconds into the flight, exploded and fell into the Atlantic about three miles from the Cape. Following that failure, three Series B missiles were launched from pads 13, 11, and 14 on August 2nd, August 28th and September 14th respectively. All of them met all of their test objectives. Another Series B missile was launched from Pad 13 on September 18th, but a turbo pump failure cut off power 84 seconds into the flight, and the ATLAS disintegrated. Another Series B missile met some of its objectives on November 17th, and the next ATLAS in the series flew the entire length of the Range (i.e., 5,500 nautical miles) on 28 November 1958. One more "B" series missile was launched into orbit from Pad 11 on December 18th to relay President Eisenhower's Christmas message to the world. (This public relations effort was called Project SCORE.) The empty ATLAS booster remained in orbit for 34 days before it reentered the atmosphere and burned up.[21]

The ATLAS flight test program was still behind schedule, but the first Series C missile was launched successfully from Pad 12 on 23 December 1958. Though the missile's data capsule was not recovered, the flight met all other test objectives. Three more Series C missiles were launched from Pad 12 during the first half of 1959, including the first ATLAS to carry a recoverable ablative nose cone. Two more Series B missiles were launched from pads 14 and 11 on January 15th and February 4th. The first exploded 109 seconds into the flight, but the second performed very well, and a support aircraft photographed the latter missile's tank and nose cone reentry. The first Series D missile was launched from Pad 13 on 14 April 1959, and two more "D" series missiles were launched from pads 14 and 13 on May 18th and June 6th. All three exploded less than 3 minutes after launch. In summary, only two of the test missiles launched in the first half of 1959 were highly successful; the other six failed to meet most of their test objectives, and their poor performance caused a 60-day slip in the ICBM's journey to operational status.[22]

After a lackluster string of launches in the first half of 1959, two very successful flights in July 1959 were heartening. The first of those, which involved a "C" Series missile launched from Pad 12 on July 21st, included the first recovery on a full-scale ATLAS nose cone. The other was a "D" series flight from Pad 11 to an impact area near Ascension on July 28th. All test objectives were met on another Series D flight from Pad 13 on August 11th, and the last Series C mission (launched from Pad 12 on 24 August 1959) ended on a high note when the missile's nose cone was recovered 5,000 miles downrange. Eight more "D" series missiles were launched in ballistic missile tests over the next four months from pads 13 and 11. Though the results of those flights were mixed, five of them were very successful, and the other three met some of their test objectives.[23]

 *ATLAS MISSILE AND SERVICE TOWER, PAD 12*

1957

Under continued pressure from an apparent "missile gap" between the U.S. and the Soviet Union, the U. S. Air Force moved quickly to activate the ATLAS as Weapon System 107A-1 at Vandenberg Air Force Base. Months before the "D" Series proved itself at the Cape, the first operational ATLAS launch complex was completed at Vandenberg, and construction of a second operational complex was underway. Launch facilities for two squadrons of ATLAS missiles were also being built at F.E. Warren Air Force Base, Wyoming and Offutt Air Force Base, Nebraska during this period. Construction for two more ATLAS squadrons began before the end of 1959 at Fairchild Air Force Base, Washington and Forbes Air Force Base, Kansas. The Air Force accepted the ATLAS on 1 September 1959, and SAC Commander Thomas S. Power declared the missile "operational" about a week later. One of three ATLAS "D" missiles was put on alert at Vandenberg's Complex 576A shortly thereafter.[24]

The operational ATLAS "D"s stood on gantry-supported pads initially, but later on they were laid horizontally in unhardened, roofless support facilities to simplify their support requirements. Earthen roofs were introduced to provide approximately 25 pounds per square inch of blast protection for the ATLAS "E" missiles that came later. The most advanced version -- the ATLAS "F" -- sat at the bottom of an underground launch facility reinforced to withstand overpressures of 100 pounds per square inch. In addition to a mixed squadron of nine ATLAS "D," "E," and "F" missiles activated at Vandenberg Air Force Base, 12 ATLAS squadrons were activated in Wyoming, Nebraska, Washington, Kansas, Oklahoma and New Mexico between September 1960 and the end of December 1962. Though the ATLAS "D" was supposed to remain operational until 1967, all three ATLAS series were phased out between May 1964 and March 1965 as part of a general retirement of the nation's first-generation ATLAS and TITAN I ICBMs. Like the THOR, the ATLAS booster was mated to a variety of high energy upper stages over the next quarter century, and, as of this writing, it remains an important part of the U.S. space program.[25]

The 6555th, Chapter III, Section 6

The 6555th's Role in the Development of Ballistic Missiles

The TITAN Ballistic Missile Program

The TITAN I Weapon System 107A-2 program was pursued initially as insurance against the ATLAS' possible failure, but it enjoyed many technological refinements that had been deliberately left out of the ATLAS to avoid delays in the ATLAS' deployment. The TITAN I was conceived as a two-staged, liquid-fueled missile. It was ten feet in diameter and 90 feet long. The 220,000-pound missile's first stage was powered by two rocket engines rated at 150,000 pounds of thrust each. The second stage was equipped with a single 60,000-pound thrust liquid-fueled rocket engine. The Glenn L. Martin Company was responsible for the TITAN I's airframe, and the Aerojet-General Corporation provided the propulsion system. Bell Telephone Laboratories had the contract for the missile's radio-inertial guidance system, and ARMA was under contract for the all-inertial guidance system. The nose cone was developed by the AVCO Corporation.[26]

The TITAN I flight test program was divided into Series I, II and III. Twelve flights were programmed for each of the first two series, and 45 Series III flights were anticipated to complete the program. Series I flights were designed to test the TITAN's first stage and explore the problem of starting the second stage's rocket engine at altitude. On Series II flights, the second stage's guidance system was operated in conjunction with the TITAN's control system, and those flights served the additional purpose of testing the TITAN's nose cone separation mechanism. Series III flights validated the performance of the TITAN I production prototype. To save time, Series I and II tests would be run concurrently with considerable overlap in the flights.[27]

BLOCKHOUSE 16 UNDER CONSTRUCTION
November 1957

Facilities for the TITAN I program were supposed to be developed on a high priority basis, but contractor and labor relations problems delayed completion of the facilities for several months beyond the original target dates. Contracts for the Cape's TITAN I facilities were awarded to the MacDonald Construction Company on 30 January 1957, but the work had to be taken over by the Macco Company following MacDonald's default on the contracts in April 1957. A strike and other labor problems delayed construction through the spring of 1957, but a temporary injunction against picketing brought 80 to 90 percent of the contractor's work force back to the job in June 1957. Despite those delays, the TITAN's assembly buildings were ready for functional tests by the summer of 1958, and the contractor shifted to around-the-clock operations in September to get the first Titan I Complex (Number 15) ready for use by the end of November 1958. Complex 16 was almost finished by the end of the year, and complexes 19 and 20 were finished in 1959.[28]

COMPLEX 20
December
1959

The first TITAN I arrived at Cape Canaveral on 19 November 1958. That missile was erected on Complex 15 on November 23rd, but it had to be sent back to Martin's factory a week later for an oxidizer line replacement. Fortunately the problem was not serious, and the TITAN flight test program was soon off to a good start. The first four TITAN I test missiles came from a single lot -- Lot A -- and they were launched from Complex 15 on 6 February, 25 February, 3 April and 4 May 1959. Isolated performance discrepancies were noted on the flights on February 25th and May 4th, but all four missiles met virtually all of their test objectives. Encouraged by those results, the contractor looked forward to the first flights of Lot B and Lot C missiles.[29]

BLOCKHOUSE 19
INTERIOR
1961

TITAN MISSILE ASSEMBLY BUILDING U
Cape Canaveral, March 1963

FIRST TITAN I LAUNCH FROM PAD 15
6 February 1959

Unfortunately, the next two launches failed dramatically. The first Lot B missile was expected to demonstrate the TITAN I's flight stability on August 14th, but, as the missile built up thrust prior to its lift-off from Complex 19, its tie-down bolts exploded early, and one of the umbilicals generated a "no-go" signal to the ground support equipment's flight controls as the missile lifted off the pad prematurely. The "no-go" signal prompted an automatic engine kill signal from the flight controls, and the TITAN lost all thrust. The missile fell back through the launcher ring and exploded, and the umbilical tower was damaged in the ensuing fire. The first Lot C launch took place at Complex 16 on 12 December 1959, and it was equally discouraging. The missile's first stage destruct package ruptured the fuel tank about four seconds after launch, and the second stage fell back on the pad and exploded. Though complexes 19 and 16 were returned to service for launches in February 1960, the next Lot C missile exploded 52 seconds after its lift-off from Complex 16 on February 5th, and two other Lot C missiles experienced second stage problems during their flights downrange on 8 March and 8 April 1960. On the other hand, the second (and last) Lot B missile met all of its test objectives on 2 February 1960, and the last Lot C missile landed in the Ascension impact area as planned on April 28th. Six Lot G missile flights to the Ascension area were also completed successfully between February 24th and the end of June 1960. Like the THOR and ATLAS before it, the TITAN I's successes soon eclipsed its failures.[30]

TITAN MISSILE EXPLOSION ON PAD 19
14 August 1959

TITAN LAUNCH FAILURE ON PAD 16
12 December 1959

TITAN I LOT C LAUNCH FROM PAD 16
28 April 1960

By the time the 6555th assumed responsibility for the technical management of the Air Force's ballistic missiles, the THOR and ATLAS programs had achieved operational status, six TITAN I test missiles had been launched from the Cape, and construction was underway on complexes 31 and 32 to support America's second-generation ICBM, the MINUTEMAN. Nevertheless, much work remained to be done, and the 6555th Test Wing (Development) had separate project test divisions and operations divisions for THOR, ATLAS and TITAN and the TS 609A (BLUE SCOUT). The Wing also had separate project test divisions for MINUTEMAN and the MIDAS satellite program, and it had an operations division for the air-breathing MACE missile program. The project divisions were grouped under the Director of Tests, who exercised on-the-spot technical supervision of contractor-conducted missile tests. The operations divisions were organized under the Director of Operations, who was charged with providing a blue suit launch capability for missile and space programs. Under the 6555th's Director of Support, there were other divisions for engineering, instrumentation, plans and requirements, facilities, materiel and inspection. Those divisions provided an Air Force test and evaluation capability for missiles and space vehicles.[31]

The 6555th, Chapter III, Section 7

The 6555th's Role in the Development of Ballistic Missiles

Organization, Resources and Activities in the 1960s

Following the 6555th's redesignation on 21 December 1959, Colonel Henry H. Eichel became the 6555th Test Wing's first commander, and Lieutenant Colonel Harry C. Swan served as the Director for Support. Lieutenant Colonel John A. Simmons, Jr. left the TITAN Operations Division to become the 6555th's Director of Operations on 4 January 1960, and Lieutenant Colonel Erwin A. Meyer, Jr. moved from his position as Chief of the ATLAS Project Division to succeed Lieutenant Colonel Edmund A. Wright, Jr. as the Wing's Director of Tests on the same date. The following officers were in charge of the divisions listed below:[32]

- Lt. Colonel Robert R. Hull - ATLAS Project Division
- Lt. Colonel Edmund E. Novotny - TITAN Project Division
- Lt. Colonel Thomas W. Morgan - (THOR) Space Projects Division
- Lt. Colonel Gene R. Swant - MINUTEMAN Project Division
- Major Arnold N. Good - MIDAS Project Division
- Lt. Colonel Jesse G. Henry - TS 609A Systems Project Division
- Major Richard B. Minor - ATLAS Operations Division
- Major Harold J. Stocks - TITAN Operations Division
- Major Howard M. Sloan - TS 609A Operations Division
- Major Abbott L. Taylor - MACE Operations Division
- Major Earl W. Anderson - THOR Operations Division
- Major Robert R. Swantz - Instrumentation Division
- Lt. Colonel Prentice B. Peabody - Plans & Requirements
- Major William F. Sandusky - Facilities Division
- Captain Russell E. Selby - Inspection Division

Under the 6555th's concept of operations, the operations divisions were manned with "test-oriented ARDC personnel" to: 1) accelerate the development of a blue suit launch capability, 2) pick up weapon system deficiencies that the contractor or less seasoned troops might miss and 3) provide trained cadres for SAC's missile operations at Vandenberg in exchange for lower grade and relatively inexperienced Air Force technicians. Blue suit training followed the traditions established during the winged missile era: when a new ballistic missile program arrived at the Cape, Air Force engineers and technicians were integrated into the contractor's work force for individual training on a non-interference basis. This training normally occurred during the first year of the program, when the contractor principal contact with the 6555th was through the Wing's project divisions. The length of individual training programs varied, but the trainees were not withdrawn and organized into launch teams until the Wing Commander considered the action appropriate. The launch teams were then reintegrated into the program to polish their skills (under the contractor's supervision, initially), and then they were assigned to a missile complex to launch a specified number of test missiles. After several successful launches, the trained cadre was transferred to SAC, and more untrained troops entered the training pipeline. A separate space launch capability for Air Force, NASA, and the Advanced Research Projects Agency (ARPA) was also maintained by the 6555th and its contractors, and the Directorate of Support designed missile tests, evaluated support requirements and checked out missile and blockhouse instrumentation.[33]

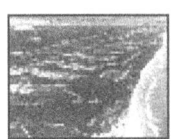

"MISSILE ROW" - CAPE CANAVERAL

1964

The 6555th's ballistic missile and space booster property holdings at the Cape were considerable. At the beginning of 1960, complexes 11 through 14 were configured for ATLAS "D" missiles, but Complex 14 also accommodated ATLAS boosters used in the MIDAS satellite program and Project MERCURY.* The ATLAS Project Division had all ATLAS complexes assigned to its jurisdiction, including Complex 36, which was under construction in 1960 for the ATLAS/CENTAUR program. All or part of hangars J, K, N, H and F were operated by ATLAS contractors, so those facilities came under the Division's supervision as well. Two missile control facilities at Cape Canaveral and San Salvador and two offices at the Cape and Patrick Air Force Base rounded out the Division's real estate holdings.[34]

MAP - CAPE CANAVERAL AIR FORCE STATION

The TITAN Project Division had jurisdiction over four TITAN complexes (e.g., 15, 16, 19 and 20), a radio-guidance site and laboratory, an all-inertial guidance lab, hangars T and U, and a reentry vehicle hangar. The TITAN project was administered from the Cape's Engineering and Administration (E&A) Building, and an office was also maintained in the AFMTC Headquarters Building at Patrick. Rounding out the old and the new, the 6555th still had jurisdiction over Complex 17, its two launch pads and blockhouse, Complex 18 and Pad 18B -- all of which were transferred to the Space Projects Division and the TS 609A Systems Project Division shortly before the THOR Operations Division's demise in April 1960. Aside from silo liners and some mobile facilities, all the Cape's MINUTEMAN facilities were completed by the end of 1960, and the 6555th had jurisdiction over them. They included two blockhouses, launch pads 31A and 32A, silos 31B and 32B, two missile assembly buildings (AB and

AC), two engine receiving and inspection buildings, a propellant inspection building, two engine storage buildings, one assembly and support building, one missile storage building and annexes in hangars I and N.[35]

At the beginning of 1960, airmen assigned to the ATLAS Operations Division were working for Convair on ATLAS ground and flight tests as part of the 6555th's on-the-job training program. As this individual training continued, Convair launched 18 ATLAS "D" and six ATLAS "E" test missiles from the Cape between 6 January 1960 and 25 March 1961. Following the 6555th's internal reorganization on 17 April 1961, the ATLAS Project Division was divided into the ATLAS Weapons Branch and the ATLAS Booster Branch. The ATLAS Operations Division was integrated into the ATLAS Weapons Branch as one of three sections (e.g., systems, requirements, and operations). Major Samuel S. McClure became the Chief of the Operations Section under Lieutenant Colonel Arnold N. Good, who was assigned as ATLAS Weapons Branch Chief. By June 1st, three Operations Section personnel were working at ARMA's guidance laboratory, and the rest of the Section's airmen had replaced contractor technicians at Complex 11 to turn that facility into a military operation. Though that transformation was not completed in 1961, the Operations Section participated in five ATLAS launches from Complex 11 in the last half of 1961, and airmen/technicians completed most of the checkout and launch items required on two of those flights. A total of 15 "E" Series E and four "F" Series missiles were launched from complexes 11 and 13 during 1961.[36]

ATLAS "F" BACKING TO ERECTOR, PAD 11

August 1962

Following Lieutenant Colonel McClure's transfer to the Martin Company's offices in Denver, Colorado in January 1962, Major J. F. Pierce became the Chief of the Operations Section. The ATLAS Weapons Branch remained under Lieutenant Colonel Arnold N. Good, but the systems and requirements sections were combined into a single Systems & Requirements Section under Major J. D. Edgington. The Branch had 14 officers, 93 airmen and eight civilians assigned to its various activities at the beginning of 1962, but the total complement grew to 142 as more airmen were assigned to the unit over the next six months. Contractor employees could still be found on Complex 11 during that period, but they constituted only about 25 percent of the work force in June 1962, and they represented only a handful of people six months later. As on-the-job training continued, only one ATLAS "F" missile was launched from Complex 11 during the first six months of 1962, but a major milestone was passed when an all-military launch crew launched its first ATLAS "F" from Complex 11 on 13 August 1962. The flight was very successful, as were the flights of four more ATLAS "F" missiles launched by Major Pierce's troops between 12 September and 6 December 1962. The flights concluded the ATLAS ballistic missile flight test program at Cape Canaveral.[37]

BLOCKHOUSE 11 CONTROL BOARD
Before 13 August 1962 Launch

ERECTION AND MATING OF ATLAS NOSE CONE AT PAD 11

August 1962

FIRST "BLUE SUIT" LAUNCH OF AN ATLAS "F" FROM PAD 11

13 August 1962

The ATLAS Weapons Division and its Operations Branch launched one ATLAS "E" and four ATLAS "F" missiles for the Advanced Ballistic Reentry System (ABRES) program between 1 March 1963 and 2 April 1964. The Division also provided an on-the-job ATLAS training program for new Air Force personnel as old troops moved to other assignments, retired, or separated from the service. Following the cancellation of the ABRES program at the Cape in the summer of 1964, the Operations Branch removed ground equipment and prepared and shipped two ATLAS "F" missiles to Norton Air Force Base before the end of the year. All remaining ATLAS Weapons Division personnel had been reassigned to other positions in the 6555th or to positions elsewhere in the Air Force by the middle of 1965.[38]

The 6555th also began developing a military launch capability for the TITAN I ballistic missile program at the Cape in 1959. By the spring of 1960, the TITAN Operations Division had completed about 50 percent of the training needed to form an all-military TITAN launch crew, and many of its airmen were working with the Martin Company as members of the contractor's TITAN firing teams. The Division's strength increased to 53 officers and men in the summer of 1960, but the Division still lacked an established plan of operation, and crowded working conditions coupled with deteriorating morale began to have a negative affect on the unit's performance. Fortunately, relief was on the way: in accordance with the 6555th's reorganization in April 1961, the TITAN Operations Division was merged with the TITAN Project Division to become the TITAN Weapons Branch. The new branch pooled the resources of both divisions, including four civilian engineers and four civilian administrative assistants. Though the Branch still had fewer officers and airmen than it needed to do its job (e.g., 18 officers and 87 airmen assigned, versus 22 officers and 108 airmen required to accomplish the mission), it was soon on its way to achieving a blue suit launch capability.[39]

ERECTION OF FIRST "BLUE SUIT" TITAN I MISSILE

October 1961

As Martin continued to launch TITANs from complexes 19 and 20, the TITAN Operations Section fielded its first all-military launch crew in November 1961. The blue suit crew launched its first TITAN I missile from Complex 20 on 21 November 1961, and that flight met all of its test objectives. The blue suit crew launched its second TITAN I from Complex 20 on December 14th, and that flight was equally successful. Unfortunately, the TITAN I test program was near its end, and Martin launched its last TITAN I test missile from Complex 19 on 29 January 1962. Within a few weeks, complexes 19 and 20 were transferred from the TITAN Weapons Branch to the THOR/TITAN Space Branch where they would be converted to support other TITAN space missions. In the near term, the TITAN II's flight test program on complexes 15 and 16 commanded Martin's attention, but space missions were already on the horizon. Martin launched its first TITAN II from Complex 16 on 16 March 1962, and the flight was highly successful.[40]

FIRST "BLUE SUIT" LAUNCH OF A TITAN I MISSILE FROM PAD 20

21 November 1961

FIRST TITAN II LAUNCH FROM PAD 16

16 March 1962

As the contractor's launches continued on complexes 15 and 16, the TITAN Operations Section shifted its focus to the TITAN II on Complex 15. The Section's people received two months of formal training at Martin's TITAN plant in Denver during the first half of 1962, and they continued their on-the-job training at Cape Canaveral. Martin launched its second TITAN II from Complex 15 on 7 June 1962, and it recorded two more successful test flights from complexes 15 and 16 on July 11th and July 25th. The Operations Branch's participation in TITAN II launches remained somewhat limited during this period, but its involvement increased significantly during three test flights on 12 September, 26 October and 19 December 1962. Finally, on 6 February 1963, the TITAN Weapons Division recorded its first blue suit launch of a TITAN II test missile. Most test objectives were met on that flight, and the second all-military TITAN II launch on March 21st was even more successful. While the third blue suit TITAN II launch failed to meet any of its test objectives on April 19th, the Operations Branch's second shift launch crew completed their TITAN II training on 21 August 1963 with a highly successful test flight from Complex 15. The Operations Branch also participated in two contractor-led TITAN II operations from

Complex 15 in November and December 1963. Both flights were successful.[41]

Four more TITAN II test flights were launched from Complex 15 in 1964 before the missile's R&D program was concluded at Cape Canaveral. Two of the flights, which were launched on 15 January and 26 February 1964, met some of their test objectives. The other two test flights, on 23 March and 9 April 1964, met all of their objectives. (Since the Operations Branch and the Martin Company each contributed half of the launch team for those operations, they each deserved half the credit for the flights.) Following the last flight in April, complexes 15 and 16 were placed in standby status while the rest of the TITAN's facilities were reassigned to other programs at the Cape. The TITAN Weapons Division was discontinued on 30 June 1964, and its personnel were reassigned to other divisions.[42]

The 6555th, Chapter III, Section 8

The 6555th's Role in the Development of Ballistic Missiles

The MINUTEMAN Ballistic Missile Test Program

The MINUTEMAN research and development program was the Air Force's last intercontinental ballistic missile effort at Cape Canaveral, but it involved the 6555th for more than a decade. The MINUTEMAN Project Division was activated on 1 January 1959 under the Air Force Ballistic Missile Division's Assistant Commander for Missile Tests. Major Gene R. Swant was assigned as Chief of the MINUTEMAN Project Division, and he continued in that capacity as a lieutenant colonel after the Division was assigned to the 6555th Test Wing (Development) on 21 December 1959. Unlike the 6555th's operations divisions, the project divisions were essentially liaison agencies, and Lieutenant Colonel Swant only had about a dozen officers and clerical assistants at the Division's offices at Patrick and Cape Canaveral. As MINUTEMAN facilities were completed at the Cape, however, missile contractor activity increased dramatically. An inert MINUTEMAN I missile was processed along with 90 percent of its support equipment in the spring of 1960 to insure dimensional compatibility between the Cape's MINUTEMAN facilities and future "live" missiles. Another inert missile (equipped with electrical components to test the facilities' electronic compatibility) was assembled and tested at the Cape in October and November 1960. Last-minute construction, equipment installation and launch pad preparations also required an around-the-clock effort from Boeing toward the end of 1960 to get the facility ready for the first MINUTEMAN I launch in early 1961.[43]

POSITIONING MINUTEMAN I ON TEST STAND, PAD 31

11 August 1960

The 6555th activated its MINUTEMAN Operations Division in July 1960, but, pending coordination of the Division's mission with higher headquarters, it felt compelled to restrict the unit's manpower to a division chief (i.e., Lieutenant Colonel Good), a lieutenant and one chief master sergeant. On 17 April 1961, the MINUTEMAN Operations Division and the MINUTEMAN Project Division were combined to form the MINUTEMAN Weapons Branch under Lieutenant Colonel Swant. Major J. J. DeJonghe and Major L. W. Sinclaire were chosen to lead the Branch's Requirements Section and System Section in April, and Major A. L. Taylor became the Chief of the Operations Section on 29 May 1961. Though the combined strength of the two units yielded only 11 officers, 18 airmen and 4 civilians initially, the complement of airmen tripled over the next eight months. Many of those troops were sent to technical schools to study the MINUTEMAN's various systems before the end of 1961.[44]

Photo Not Available

BLOCKHOUSE 31 (LEFT) AND 32 (RIGHT) UNDER CONSTRUCTION
March 1960

As work on Silos 31 and 32 neared completion, the first MINUTEMAN I test missile was launched from Pad 31 on 1 February 1961. The flight was highly successful, and it set a record for being the first launch operation in which all stages of a multi-staged missile were tested on the very first test flight of an R&D program. Successes alternated with failures when the second and fourth MINUTEMAN I missiles were destroyed during their flights from Pad 31 and Silo 32 on 19 May and 30 August 1961, but two other MINUTEMAN flights were launched from Silo 32 and Silo 31 before the end of 1961, and they met most of their test objectives. Apart from one flight failure in April 1962, Boeing had a string of five successful flights from Silo 31 between 5 January and 9 March 1962, and the Cape recorded four more successful test flights from Silo 32 in May and June 1962. (The latter included the first all-military launch of a MINUTEMAN I missile on June 29th.) After a bad start, test results in the last half of 1962 were also somewhat mixed: two MINUTEMAN I test missiles destroyed themselves during test flights in July and August 1962, and another MINUTEMAN I had to be destroyed by the Range Safety Officer approximately eight seconds after launch on October 17th. Five successful test flights were recorded in September, November and December 1962, and the year's operations were capped by a partially successful flight from Silo 32 on December 20th.[45]

MINUTEMAN SILO 32 UNDER CONSTRUCTION
March 1960

VIEW OF BLOCKHOUSE 32 AREA

May 1960

BLOCKHOUSE 31

February 1961

The 6555th, Chapter III, Section 8, The MINUTEMAN Ballistic Missile Test Program

FIRST MINUTEMAN I LAUNCH FROM PAD 31

1 February 1961

INTERIOR OF BLOCKHOUSE 32

December 1962

BLOCKHOUSE 32 CONSOLE DURING COUNTDOWN
December 1962

Before the first blue suit MINUTEMAN I launch in June 1962, military personnel from the MINUTEMAN Weapons Branch's Operations Section attended factory training courses and worked with Boeing, Autonetics and AVCO to learn the various facets of missile preparation, ground testing, launch operations and silo refurbishment. The Section's officers and airmen also participated in varying degrees on all the MINUTEMAN launch operations during the first half of 1962, and they redoubled their efforts on Complex 32 to prepare Silo 32 for the next block of MINUTEMAN missiles. As part of the Wing's reorganization in the summer of 1962, the MINUTEMAN Weapons Branch became a division and the Operations Section became a branch, but the mission did not change. Of the nine MINUTEMAN I flights registered in the last half of 1962, four were all-military launch operations from Silo 32. The contractor launched the other five MINUTEMAN I test missiles from Silo 31 on 12 July, 19 September, 17 October, 19 November and 14 December 1962.[46]

The MINUTEMAN Weapons Division assumed responsibility for Complex 31 in 1963, and this decision prompted the assignment of 31 additional airmen, five SAC officers and 17 SAC airmen to assist the Division with its expanded operation. Unfortunately, most of the new personnel had no previous MINUTEMAN training, so the Operations Branch had to conduct a series of orientation courses on the MINUTEMAN weapon system and R&D testing procedures before the new arrivals were allowed to move on to the on-the-job training phase on their instruction. By June 1963, the Division had 16 officers, 91 airmen and 10 civilians assigned to its various activities. Eight SAC officers and 20 SAC airmen were also attached to the Division, and all new personnel had progressed to the on-the-job phase of their training. In the meantime, Boeing launched its last four MINUTEMAN I test missiles on successful flights from Silo 31 on 7 January, 18 March, 10 April and 28 May 1963. Blue suit operations also continued from Silo 32, and five more successful MINUTEMAN I flights were launched from that facility between 23 January and 28 June 1963. The first blue suit MINUTEMAN I launch operation from Silo 31 led to a completely successful missile flight on 27 June 1963.[47]

While MINUTEMAN I launches continued at Cape Canaveral, other aspects of the MINUTEMAN program advanced elsewhere in the United States. On 28 September 1962, for example, a MINUTEMAN I missile was launched from Vandenberg Air Force Base for the first time in that base's history. The first MINUTEMAN I (model "A") flight of 10 missiles was placed on alert at Malmstrom Air Force Base, Montana on 27 October 1962, and the first full squadron of 50 MINUTEMAN I missiles was on alert at Malmstrom by the spring of 1963. The first MINUTEMAN I (model "B") missiles went on alert at Ellsworth Air Force Base, South Dakota in July 1963, and Ellsworth's 66th Strategic Missile Squadron was declared operational less than three months later. Technological improvements in the MINUTEMAN had already out-distanced its deployment by that time, and the Secretary of Defense approved a program in November 1963 to gradually replace the entire MINUTEMAN I "A" and "B" force with more powerful MINUTEMAN II missiles. By July 1964, 600 MINUTEMAN I missiles were dispersed in hardened underground launch facilities at 12 operational missile squadrons in the western United States. They constituted roughly three-fourths of a mixed force of ATLAS, TITAN and MINUTEMAN missiles, but their representation increased considerably as all but 54 of the first-generation ICBMs were retired over the next 12 months. Only 54 TITAN II missiles were retained as the MINUTEMAN force continued to expand to 1,000 MINUTEMAN I and II missiles. Malmstrom's 564th Strategic Missile Squadron completed the deployment by putting the 1000th MINUTEMAN on alert in May 1967.[48]

FIRST MINUTEMAN II AT CAPE CANAVERAL

September 1964

MINUTEMAN TRANSPORTER/ERECTOR ALIGNED WITH SILO 31

Between 1 July 1963 and 30 September 1964, the 6555th's MINUTEMAN Operations Branch launched nine missiles from Silo 31 and 10 missiles from Silo 32 to conclude the MINUTEMAN I flight test program. Fourteen of those flights met all -- or a very high percentage -- of their test objectives, and the 6555th completed a string of 12 successful launches without a single flight failure in 1964. Facilities were reconfigured for the MINUTEMAN II program during the last half of 1964, and the Operations Branch launched the first MINUTEMAN II test missile from Silo 32 on September 24th. Three additional highly successful MINUTEMAN II flights were launched from Cape Canaveral before the end of 1964, and they were followed by a string of seven near-perfect test flights from silos 31 and 32 in 1965. Only nine more missile flights were needed to conclude the MINUTEMAN II program at the Cape, so Complex 31 was put into temporary storage in the spring of 1966 pending its modification for the MINUTEMAN III program. The Operations Branch launched four MINUTEMAN II test missiles in 1966, and it launched four more in 1967. The final MINUTEMAN II was launched from the Cape on 6

February 1968. As MINUTEMAN II operations wound down, Lieutenant Colonel William E. Haynes became the MINUTEMAN Weapon Division Chief on 12 April 1966, and he was succeeded by Lieutenant Colonel Joseph M. Glasgow, Jr. on 1 January 1967. Fifteen officers, 90 airmen and seven civilians were assigned to the Division by the end of 1967, but one officer and sixteen airmen were added to the unit as the MINUTEMAN III program got underway in 1968.[49]

Though the Operations Branch dominated MINUTEMAN launch operations from the middle of 1963 onward, it would be extremely unfair to ignore the contributions made by Boeing and its MINUTEMAN sub-contractors throughout the flight test program. All MINUTEMAN test missiles were assembled at the Cape by contractor personnel and tested before they were turned over to the Air Force and transported to the silos. (As part of the procedure, the 6555th's MINUTEMAN Systems Branch assigned its own personnel to supervise the contractor's assembly and sub-system testing before Boeing transferred the missiles to the Operations Branch.) The contractor's participation in the MINUTEMAN program was thus quite extensive. At the end of 1967, for example, Boeing had 324 employees assigned to the MINUTEMAN program at Cape Canaveral, and MINUTEMAN sub-contractors provided 140 workers for their portions of the assembly and testing operation. TRW (formerly Ramo-Wooldridge) also provided more than two dozen people to support the MINUTEMAN II and MINUTEMAN III programs. Taken together, approximately 45 percent of the 6555th's entire missile contractor work force were involved in the MINUTEMAN program by the middle of 1968. MINUTEMAN contractor numbers dropped to 527 a year later, and they plunged from 401 to 137 as the MINUTEMAN III program wound down during the last half of 1970, but a comparison of those figures with military strengths in 1969 and 1970 clearly shows that contractors outnumbered MINUTEMAN Weapons Division personnel at least 4-to-1 during MINUTEMAN III operations at Cape Canaveral.[50]

The military also had an important role in MINUTEMAN operations. Once the contractors delivered the assembled MINUTEMAN to the 6555th, personnel from the MINUTEMAN Operations Branch's Pad-Silo Section drove the missile to the pad in a special vehicle known as a transporter/erector. They lowered the MINUTEMAN into the silo, installed and checked out the missile's control umbilicals, mated the guidance and instrumentation section to the missile, installed secondary ordnance and operated special test equipment required to calibrate and record silo instrumentation data. The Blockhouse Section's technicians performed pre-flight tests to insure proper control of all systems before launch, and the Blockhouse Section's officers served as MINUTEMAN test conductors. The MINUTEMAN Weapon Division's Inspection Branch monitored all phases of those operations, including the actual launch. Following lift-off, the Pad-Silo Section's people moved in to refurbish the missile suspension system and the launch tube.[51]

As preparations for the first MINUTEMAN III launch entered their final phase, Lieutenant Colonel Glasgow's tour as Chief of the MINUTEMAN Weapon Division ended, and he was relieved for reassignment on 22 July 1968. He was succeeded by Lieutenant Colonel Arthur E. Hendren, a veteran missileer and recent arrival from Vandenberg's 6595th Aerospace Test Wing. Under Lieutenant Colonel Hendren, the Operations Branch successfully launched the first MINUTEMAN III test missile from Silo

32 on 16 August 1968. That flight was followed by nine other test flights from Silo 32 and Silo 31 between 24 October 1968 and 13 March 1970.* Though four of those later MINUTEMAN III flights failed to meet their test objectives, the Operations Branch wrapped up the MINUTEMAN III R&D flight test program with three highly successful flights from Silo 32 between 3 April and 28 May 1970. When the 6555th became a Group under the 6595th Aerospace Test Wing on 1 April 1970,** the MINUTEMAN Weapon Division was renamed the MINUTEMAN Test Division, but the name change was a minor event compared to the termination of blue suit launch operations and the subsequent transfer of personnel to other agencies. Lieutenant Colonel Hendren's division reduced its manpower to 16 officers, 60 airmen and six civilians by 1 July 1970, and it got rid of its blue suit launch capability. Though three more MINUTEMAN III missiles were launched from Silo 32 on 16 September, 2 December and 14 December 1970, they were launched by Boeing for the Special Test Missile (STM) project -- a post-R&D effort to evaluate the MINUTEMAN III's performance and accuracy. (All three test flights were successful.) Following the final MINUTEMAN launch on December 14th, the MINUTEMAN Test Division continued to reduce its numbers, and only a handful of personnel were retained to complete the disposition of MINUTEMAN equipment after the Division was deactivated on 31 December 1970. The remaining personnel were reassigned to other duties, and the last of the MINUTEMAN contractors departed in 1971.[52]

FIRST MINUTEMAN III LAUNCH FROM SILO 32

16 August 1968

The 6555th's role in ballistic missile development ended with the MINUTEMAN III flight test program in 1970, but MINUTEMAN and TITAN missile tests continued under SAC and the 6595th Aerospace Test Wing at Vandenberg Air Force Base. Many improvements in those missiles and their reentry systems were tested and verified at Vandenberg and the Western Test Range, and new Air Force ballistic missile programs (e.g., PEACEKEEPER and the Small ICBM) were added to the Western Test Range's schedule in later years. The 6555th's mission, on the other hand, was focused on space launch vehicles, payloads, and support systems during the 1970s, and the Group continued to lead the way for space operations it had pioneered in the 1960s. In the next chapter, we will look at the 6555th's involvement in space activities dating back to 1959.

The 6555th

Chapter Three Footnotes

ballistic missile as a logical successor
Following Defense Department cutbacks in 1947, the Consolidated-Vultee Aircraft Corporation had been forced to close out its MX-744 ballistic missile flight test program in 1948. Consolidated's successor -- Convair -- continued limited research on ballistic missile technology with its own corporate funds.

Strategic Missiles Evaluation Committee
The 11-member panel consisted of three of Hughes Aircraft's top men (Allen E. Puckett, Simon Ramo, and Dean Wooldridge), Brigadier General (Selectee) Bernard A. Schriever (who was then Assistant for Development Planning under the Air Force Deputy Chief of Staff for Development), three professors from the California Institute of Technology (Clark B. Millikan, Charles C. Lauritsen and Louis G. Dunn), Hendrik W. Bode (from Bell Lab), George B Kistiakowski (from Harvard), Jerome B. Wiesner (from the Massachusetts Institute of Technology), and Lawrence A. Hyland (from Bendix Corporation). Von Neumann, Bode, Hyland, Kistiakowski, Millikan and Wiesner also served on the Scientific Advisory Committee, which provided the Secretary of Defense and the Secretary of the Air Force with expert advice on technological matters in general.

THOR
The THOR grew out of a meeting of the Scientific Advisory Committee in January 1955. The United Kingdom expressed an interest in the THOR in February 1955, so basing for the 1,500-mile-range missile would not be a problem. General Schriever opposed the THOR initially, fearing that it would take funding away from the ATLAS and TITAN. The Secretary of Defense thought otherwise, and President Eisenhower approved a National Security Council recommendation on 1 December 1955 assigning the THOR "joint highest national priority" with the ATLAS and TITAN.

solid propellant rocket
Although the WDD transferred most of its early solid rocket research to the Wright Air Development Center, the Division managed experiments involving the X-17 solid rocket reentry test vehicle. (The X-17 was used to establish a design for the ATLAS' reentry vehicle, and X-17 flights were launched from Cape Canaveral between May 1955 and the end of August 1957.) Toward the end of 1957, the Air Force appeared ready to develop large solid rockets for the three-staged MINUTEMAN ICBM, whereupon General Schriever requested the return of other solid rocket projects to his division.

ballistic missile components were being tested elsewhere

An ATLAS component test facility was established at Point Loma, and captive missile test facilities were set up at the Edwards Rocket Site north of Los Angeles and at Sycamore Canyon near San Diego. The first ATLAS propulsion system and component tests were conducted in June 1956, and the first ATLAS test missile was delivered to Sycamore Canyon in August 1956. The Air Force and the Douglas Aircraft Company jointly financed a static test (i.e., captive missile) firing facility for the THOR at Sacramento, California. The site was leased, with an option to buy, from Aerojet General Corporation. Douglas delivered its first THOR test missile on 26 October 1956, and that missile was launched from Cape Canaveral three months later.

Assistant Commander for Missile Tests
Unfortunately the Office of Assistant Commander for Missile Tests was discontinued on 21 December 1959, so the 6555th only picked up the Office's resources and mission -- not its lineage and honors.

ballistic missile testing
For the sake of clarity, we will limit our study to the THOR, ATLAS, TITAN I and II, and MINUTEMAN I, II, and III missile systems. No attempt will be made to compare those programs with the development of the Army's REDSTONE and JUPITER missiles or the Navy's POLARIS, though those missiles also had a rich history at the Cape. Some parallels between operations at Vandenberg Air Force Base and Cape Canaveral will be presented to illustrate other aspects of the Air Force programs, but our principal focus will remain with the 6555th and the Air Force's ballistic missile contractors

THOR-ABLE
The ABLE upper stage was a modified Aerojet-General booster rated at 7,700 pounds of thrust. The nose cones, built by General Electric and AVCO, were designed to absorb the intense heat of atmospheric reentry by shedding thin layers of their surfaces.

Vandenberg Air Force Base
In June 1956, the ATLAS Site Selection Board (appointed by Major General Schriever) recommended an inactive Army base north of Lompoc, California as the initial West Coast base for ballistic missile operations and operational personnel training. That base -- Camp Cooke -- had been used as an armor and infantry training reservation during the Second World War and the Korean War, but it had been on inactive status between the wars and after January 1953. The Air Force went ahead with plans for three ATLAS complexes and the development of a THOR operational training capability after Secretary of Defense Charles E. Wilson directed the transfer of 64,000 acres of Camp Cooke to the Air Force on 16 November 1956. The 6591st Support Squadron became the first Air Force unit to move into Camp Cooke in February 1957, and the Air Force portion of the reservation was redesignated Cooke Air Force Base on 7 June 1957. Cooke Air Force Base was transferred from ARDC to SAC on 1 January 1958, and it was redesignated Vandenberg Air Force Base in honor of General Hoyt S. Vandenberg on 4 October 1958. Though SAC owned the base, the Air Force Ballistic Missile Division remained responsible for the design and installation of ballistic missile facilities.

ATLAS' development
Convair formed Astronautics in the mid-1950s to work on the ATLAS program, and General Dynamics' Board of Directors voted $20 million to acquire land and build a complex for Astronautics in San Diego, California. Astronautics transferred its operations to the new 12-building complex in 1958. Twenty-two new buildings were added over the next three years, and Astronautics tripled its work force from 9,618 to more than 27,000 employees by the middle of 1961.

X-17 program
The Lockheed Aircraft Corporation was awarded a contract to launch 1/4, 1/2, and full scale models of the ATLAS reentry vehicle and acquire data on the atmosphere's affects on the vehicle's skin temperatures, pressure, acceleration, radiation, dislocation and ion density. The X-17 -- a 41-foot-long, three-stage solid propellant rocket weighing 10,650 pounds -- was used to boost the models to ultrasonic speeds. Three 1/4 scale, three 1/2 scale, and seven full scale models were launched successfully during the X-17's developmental phase between May 1955 and July 1956. Twenty-four more X-17s were launched as part of the project's research phase beginning in July 1956, and 18 of those met all their test objectives. The flight test portion of the X-17 project was concluded successfully with the final launch on 21 March 1957. The X-17 program's final report was released on 10 May 1957, but one additional post-program X-17 launch was conducted on 22 August 1957 to gather more data on nose cone vibration.

TITAN I
The TITAN I was also known as the XSM-68 (Xperimental Strategic Missile 68) while it was under development. Though the TITAN trailed the ATLAS by only one year, the technological differences in the two missiles were significant.

Contracts for the Cape's TITAN I facilities
Missile assembly buildings N, T and U and launch complexes 15, 16, 19, and 20 were included in those contracts.

BLUE SCOUT
The BLUE SCOUT program was initiated in June 1959 with the selection of Ford's Aeronutronic Division as the system's engineering contractor. The BLUE SCOUT and BLUE SCOUT Jr. were launched by blue suit crews from Complex 18 on a variety of space research missions, but they were never intended for operational missile units like the THOR squadrons in Great Britain or the TITAN and MINUTEMAN squadrons under SAC.

officers were in charge of the divisions
A vacancy in the Chief's position at the Engineering Division was filled by Mr. Wallace R. MacGregor pending the arrival of Lieutenant Colonel Malcolm D. Hart, who was assigned as the Division's Chief on 4 August 1960. Captain (and later Major) Harold T. Blackburn was named

Acting Chief of the Materiel Division on 14 April 1960, and he was succeeded by Major Douglas W. Londeree on 3 August 1960.

MACE Operations
For MACE activities, see Chapter II, Section 2.

Air Force technicians
The latter would be transferred to the 6555th for further on-the-job training. Ironically, the 6555th had to ask SAC to transfer 67 trained airmen from Vandenberg's 392nd Missile Training Squadron to fill vacancies in the 6555th's THOR Operations Division in early 1960 because only six of the 20 personnel assigned to the Division were qualified for THOR operations. The Division Chief, Major Anderson, visited Vandenberg Air Force Base in February 1960 to review personnel records for the proposed transfer, but lack of funding and personnel forced the Wing to abolish the THOR Operations Division on 20 April 1960. The 6555th's THOR space booster activities continued through its Space Projects Division, but THOR ballistic missile flights ended on the Eastern Test Range as of 29 February 1960. The Division's personnel were transferred to other divisions within the Directorate of Operations and the Directorate of Tests.

complexes 11 through 14
Complex 13 was converted to handle the new ATLAS "E" in 1960, and the first ATLAS "E" was launched there on 11 October 1960. Complex 11 was also converted for "E" and "F" Series launches later on.

test missiles
While the majority of those flights met some or all of their objectives, one ATLAS "D" exploded shortly after its lift-off from Complex 13 on 10 March 1960, and another ATLAS "D" exploded while it sat on the launch pad at Complex 11 on 7 April 1960. The first three ATLAS "E" flights were launched from Complex 13, and they were equally dismal: all three missiles broke up or lost power less than three minutes after lift-off on 11 October 1960, 29 November 1960, and 24 January 1961. Looking on the bright side, the last "D" Series flight (from Complex 12) was successful on 23 January 1961, and the fourth "E" series flight (from Complex 13) was successful on February 24th.

internal reorganization
The Wing's reorganization reflected an evolutionary process that had been going on at higher headquarters for some time. By 1960, the Air Force's ballistic missile and space programs had grown too large to be managed effectively under the Air Force Ballistic Missile Division (AFBMD). Lieutenant General Schriever had been the ARDC Commander since April 1959, and he recommended the transfer of AFBMD's ballistic missile functions from ARDC's Los Angeles complex to Norton Air Force Base where they could be combined with missile site activation offices under Air Materiel Command's Ballistic Missiles Center (BMC). Space programs would be retained under AFBMD in Los Angeles. The proposal was approved, but it was soon

overshadowed by a more comprehensive reorganization of the Air Force missile and space effort. By the beginning of 1961, Soviet space accomplishments and NASA's march to absorb U.S. military space programs compelled the Department of Defense to forge a new, strong military space effort. In March 1961, Air Force Secretary Eugene M. Zuckert announced that the Air Force would be given primary responsibility for the military space mission. On 1 April 1961, Air Force Systems Command and Air Force Logistics Command were created (along with the Office of Aerospace Research in Washington D.C.) to replace ARDC and operations. Air Materiel Command and the ballistic missile and space agencies at Norton and Los Angeles were reorganized into the Ballistic Systems Division (BSD) and the Space System Division (SSD) under Air Force Systems Command (AFSC). The 6555th reorganized its Directorate of Tests, Directorate of Operations and Directorate of Support into the Ballistic Missiles Division, the Space Programs Division, and the Technical Support Division on April 17th to reflect that higher headquarters reorganization (i.e., the 6555th now served two masters under AFSC.) The 6555th's old project and operations divisions were reorganized into booster branches and weapons branches. The Wing was redesignated the 6555th Aerospace Test Wing under BSD on 25 October 1961, and it was transferred to Space Systems Division on 1 July 1963.

launched from Complex 11
The missile exploded less than one second after lift-off on April 9th. The complex sustained minor structural and major electrical damage as a result of that failure, but repairs to damaged ground equipment gave the Branch's military personnel a chance to study support systems "inside and out" while they assisted the contractor with Complex 11's refurbishment.

ATLAS Weapons Division
The ATLAS Weapons Branch was elevated to division status during the summer of 1962, though no precise date is given in the unit histories. The Systems & Requirements Section and the Operations Section both became branches when the ATLAS Weapons Branch became the ATLAS Weapons Division, and the reader may assume that all three actions occurred on or about the same date. The Systems & Requirements Branch was renamed the Test Support Branch in 1963.

new Air Force personnel
Though the Division and the Operations Branch both lost their chiefs to other AFSC Divisions in July 1964, Major Harry B. Cadwell succeeded Lieutenant Colonel Good, and Captain Daniel G. Vaughan replaced Major Pierce on 30 October 1964.

TITAN Operations Division
Lieutenant Colonel John A. Simmons was the TITAN Operations Division Chief initially, but he became the 6555th's Director of Operations on 4 January 1960, and Major Harold J. Stocks succeeded him as Chief on the same date.

plan of operation
Some of the Division's airmen participated in the ground testing and launch of Martin's first

"advanced J" TITAN from Complex 20 in late December 1960, and the number of airmen gradually increased on the contractor firing teams that launched three other "J"s on 10 February, 20 February and 23 May 1961. Nevertheless, the absence of an operations plan made an effective all-military launch capability impossible. The Martin Company was also having some trouble with its test flights during the period. The first of the Lot "J" missiles was launched from Complex 20 on 1 July 1960, but the missile had to be destroyed by the Range Safety Officer 11 seconds after lift-off. Two other Lot "J" missiles were launched from complexes 20 and 19 on July 28th and August 10th, but they both failed to meet their test objectives. A "G" Lot TITAN was launched from Complex 15 on September 29th, and it did not meet its primary flight test objective either. Fortune began to smile again with the fourth Lot "J" missile, which was launched from Complex 20 on 30 August 1960. That flight met its primary test objectives, and two more "J" missile flights (from complexes 20 and 19) were completely successful on 7 and 24 October 1960.

TITAN Weapons Branch
Lieutenant Colonel Edmund E. Novotny, who had served as the TITAN Project Division Chief, became the TITAN Weapons Branch Chief. Major Stocks became Novotny's Operations Section Chief.

continued to launch TITANs
In addition to the "J" missile flights mentioned earlier, the Martin Company launched nine "J"s between 20 January and 25 October 1961. The contractor also launched the first six Lot "M" all-inertially guided TITANs in 1961.

Operations Branch's
The TITAN Weapons Branch and its sections were upgraded to a division and three branches during this period, but the unit's operations continued unchanged. Major Fountain M. Hutchison succeeded Lieutenant Colonel Novotny as the Division Chief in August 1962, but Major Stocks continued as the Operations Branch Chief until his reassignment (as a lieutenant colonel) in July 1963. Stocks was succeeded by Lieutenant Colonel Harold T. Blackburn as Operations Branch Chief.

successful
Both missiles were launched by the Operations Sections "first shift" personnel from Complex 15. The first shift also launched the third blue suit TITAN II operation on 19 April 1963.

MINUTEMAN research and development program
Interest in a second-generation, solid-propellant ICBM rose dramatically after the Soviet Union orbited Sputnik I in October 1957, and solid rocket research was transferred to the Air Force Ballistic Missile Division in December 1957 with that interest in mind. Though the MINUTEMAN did not receive the Air Force's A-1 priority until 4 September 1959, it capitalized on "lessons learned" during the ATLAS and TITAN R&D programs as well as the Navy's interest in the solid-fueled POLARIS ballistic missile. The Boeing Aircraft Company became the prime contractor for

the MINUTEMAN, providing system integration, installation, and acceptance testing of the missile and its support systems at 1,000 launch facilities in the western U.S. Thiokol, Aerojet-General and Hercules were sub-contractors for the three-staged ICBM's propulsion system. AVCO and General Electric had contracts for the reentry vehicles, and Autonetics (a division of North American Rockwell) was in charge of the guidance system. The 74-inch diameter MINUTEMAN was fielded in three versions: 1) the MINUTEMAN I, which measured 55.9 feet in length and weighed approximately 65,000 pounds, 2) the 59.8-foot-long MINUTEMAN II, which weighed 70,000 pounds and 3) the MINUTEMAN III, which had the same dimensions as the MINUTEMAN II, weighed 76,000 pounds, and offered a multiple reentry vehicle capability.

missile contractor activity increased
Though no mention of the Division's participation appears in the histories, it is fair to assume that the compatibility tests were run by Boeing Aircraft Company and sub-contractor personnel with Lieutenant Colonel Swant's airframe, operations and propulsion system project officers looking on.

Pad 31
Only the first three MINUTEMAN I test missiles were surface-launched from Pad 31. Later MINUTEMAN test flights from Cape Canaveral were launched from underground silos 31 and 32.

flight failure in April 1962
A MINUTEMAN I flight from Silo 32 on 24 April 1962 ended about 10 nautical miles from the Cape after the missile's first stage motor failed about 20 seconds after launch.

silo refurbishment
Though the MINUTEMAN silo liners were designed to provide considerable flame protection, thereby decreasing the time it took to repair a silo after a missile launch, missile exhausts inevitably scorched the launch tubes. A MINUTEMAN silo always required some refurbishment after a launch.

reorganization
Major Robert C. Buckley became Chief of the Systems Branch on July 2nd, and Lieutenant Colonel William L. Dienst succeeded Lieutenant Colonel Swant as the Division Chief on 25 August 1962, but those changes were prompted by the cyclical turnover of personnel which typified Air Force assignments -- not the MINUTEMAN Weapons Branch's upgrade to Division status. Major John W. Planinac succeeded Major Taylor as Chief of the Operations Branch on 8 October 1963.

1964
As flight tests continued, the MINUTEMAN Weapons Division became known as the MINUTEMAN Weapon Division in 1964. Most of the SAC personnel attached to the MINUTEMAN Weapon Division were reassigned during the first half of 1965, but 21 additional airmen were assigned to the Division during the same period. Lieutenant Colonel Planinac and

Major James D. Duval, Jr. were designated as the MINUTEMAN Weapon Division Chief and the Operations Branch Chief (respectively) on 4 June 1965. By the end of 1965, the Division had 17 officers, 121 airmen, and seven civilians assigned to its various activities. One SAC officer and one SAC enlisted man were also attached to the Division.

contractor's participation
During the first half of 1968, the MINUTEMAN contractor work force rose to 482 Boeing workers. Autonetics increased its staff from 74 to 83 employees during that period, and General Electric's presence on the Range grew from 49 employees to 61 employees by the end of June 1968. AVCO decreased its representation from 10 to 6 employees during the first half of 1968, and Hercules cut its three-member staff in half (at least on paper). Thiokol dropped its two-worker operation completely, but Aerojet-General maintained two employees to support MINUTEMAN III operations at the Cape.

Silo 31
Four of the nine flights were launched from Silo 31 between 26 March and 24 September 1969. Silo 31 was deactivated following its last MINUTEMAN III launch operation on 23 September 1969.

6555th became a Group
During the 1960s, the 6555th reported initially to the Ballistic Systems Division, subsequently to the Space Systems Division, and finally to a successor that consolidated both operations under one headquarters: the Space and Missile Systems Organization (SAMSO). Unfortunately, by the end of the 1960s, a decline in Air Force missile and space operations -- particularly at Cape Canaveral -- prompted Air Force Systems Command to consolidate its launch activities and the Western Test Range's operations under a single headquarters below SAMSO. Thanks to the reorganization on 1 April 1970, the Air Force Western Test Range Headquarters at Vandenberg was inactivated and its resources were transferred to a new organization along with the 6595th Aerospace Test Wing. This new organization was the Headquarters, Space and Missile Test Center (SAMTEC), and it absorbed the 6555th by placing it under the 6595th Aerospace Test Wing. Thus, the 6555th dropped two levels in Systems Command's hierarchy as a result of the reorganization on April 1st.

The 6555th

Chapter Three Endnotes

1. General Dynamics/Astronautics, "History of General Dynamics/Astronautics," o/a June 1961, p. 2.

2. Stanley, An Air Force Command for R&D, p. 22; Neufeld, Jacob, Ballistic Missiles in the United States Air Force, 1945-1960, Office of Air Force History, 1990, pp. 93, 98, 99, 117.

3. Stanley, An Air Force Command for R&D, p. 24; General Dynamics/ Astronautics, "History," p. 3.

4. Douglas Report SM-41860, "The Thor History," May 1963, pp. 1, 3; Neufeld, Ballistic Missiles, pp. 134, 136, 143, 144, 147, 167, 182; Stanley, An Air Force Command for R&D, p. 27.

5. History of the AFBMD Field Office, 1 May 1956 - 31 October 1958, pp. 1, 2, 3, 11; History of the AFBMD Field Office, 1 November 1958 - 31 May 1959, p.1; AFMTC History, 1 July - 31 December 1959, p. 55; Letter, Major General Schriever to Major General Don N. Yates, AFMTC Commander, 24 July 1956; General Dynamics/Astronautics, "History," p. 3; Douglas Report SM-41860, "Thor History," pp. 3, 4; Crespino, "Launches," pp. 59, 60.

6. Marven R. Whipple, "List of Commanders," p. 2-2; History of the AFBMD Field Office, 1 May 1956 - 31 October 1958, p. 6; ESMC History Office, "List of 6555th ASTG Commanders, 1964 - 1981," undated; General Order Number 238, HQ ARDC, 14 December 1959.

7. AFMTC History, 1 July - 31 December 1955, p.45; AFMTC History, 1 July - 31 December 1956, pp. 61, 62; AFMTC History, 1 July - 31 December 1957, pp. 47, 48; AFMTC History, 1 July - 31 December 1958, pp. 51, 52; AFMTC History, 1 July - 31 December 1959, pp. 54, 55; History of the Headquarters, 6555th Test Wing (Development), 21 December 1959 - 31 March 1960, p. 1.

8. Crespino, "Launches," pp. 6-9 and introductory chart; AFMTC History, 1 January - 30 Jun 1956, pp. 105-107.

9. AFMTC History, 1 January - 30 June 1956, pp. 230, 231.

10. Ibid., pp. 232-235; AFETR History, Volume II, 1964, p. 95; ESMC History, 1 October 1989 - 30 September 1990, pp. 206, 207; Interview, Mr Frank Mann, CSR Plans Office, with Mark C Cleary, 19 August 1991.

11. ESMC History, 1 October 1989 - 30 September 1990, pp. 227, 228; AFMTC History, 1 January - 30 June 1956, pp. 243, 244.

12. AFMTC History, 1 January - 30 June 1956, pp. 246, 247; ESMC History, 1 October 1989 - 30 September 1990, pp. 22, 247-249.

13. AFMTC History, 1 January - 30 June 1959, pp. 141-143.

14. AFMTC History, 1 July - 31 December 1955, pp. 205-207, 325; AFMTC History, 1 January - 30 June 1956 p. 207; AFMTC History, 1 January - 30 June 1957, pp. 209-211, 215.

15. Report SM-41860, "Thor History," p. 4; AFMTC History, 1 January - 30 June 1957, pp. 211, 212; AFMTC History, 1 July - 31 December 1957, pp. 183- 185.

16. AFMTC History, 1 January - 30 June 1958, pp. 155-157; AFMTC History, 1 July - 31 December 1958, pp. 177-182; AFMTC History, 1 January - 30 June 1959, pp. 167-171; AFMTC History, 1 July - 31 December 1959, p. 177.

17. Report SM-41860, "Thor History," pp. 8, 9; ARDC History, Volume III, 1 July - 31 December 1960, pp. 3, 8; Headquarters SAMTEC History Office, "SAMTEC Chronology," o/a June 1990, pp. 1, 2; Neufeld, Ballistic Missiles, pp. 186, 232.

18. General Dynamics/Astronautics, "History," pp. 3, 4; AFMTC History, 1 January - 30 June 1955, pp. 345-348; AFMTC History, 1 January - 30 June 1956, pp. 191, 192, 198; AFMTC History, 1 July - 31 December 1956, pp. 190- 192; AFMTC History, 1 January - 30 June 1957, pp. 195, 196 and 199; AFMTC History, 1 July - 31 December 1957, pp. 178, 179.

19. AFMTC History, 1 January - 30 June 1957, pp. 199, 201; AFMTC History, 1 January - 30 June 1956, p. 305; AFMTC History, 1 July - 31 December 1956, pp. 330, 331.

20. AFMTC History, 1 July - 31 December 1957, p. 180; AFMTC History, 1 January - 30 June 1958, pp. 152, 153.

21. AFMTC History, 1 July - 31 December 1958, pp. 168-171.

22. Ibid., p. 171; AFMTC History, 1 January - 30 June 1959, pp. 160-162; ARDC History, 1 January - 30 June 1959, p. II-59.

23. AFMTC History, 1 July - 31 December 1959, pp. 167-171.

24. Neufeld, Ballistic Missiles, pp. 191, 208; ARDC History, p. II-20, II- 21.

25. Neufeld, Ballistic Missiles, pp. 192, 233-235, 238.

26. AFMTC History, 1 January - 30 June 1956, pp. 200, 201; AFMTC History, 1 January - 30 June 1957, p.205.

27. AFMTC History, 1 January - 30 June 1957, pp. 205, 206.

28. AFMTC History, 1 January - 30 June 1957, p. 207; AFMTC History, 1 January - 30 June 1958, p. 206; AFMTC History, 1 July - 31 December 1958, pp. 173, 174; AFMTC History, 1 January - 30 June 1959, p. 244; AFMTC History, 1 July - 31 December 1959, p. 265.

29. AFMTC History, 1 July - 31 December 1958, pp. 175, 176; AFMTC History, 1 January - 30 June 1959, pp. 164, 165.

30. AFMTC History, 1 July - 31 December 1959, pp. 174, 175; Marven R. Whipple, "Index of Missile Launchings by Missile Program, July 1950 - June 1960," 15 December 1960, pp. 12-2, 12-3.

31. AFMTC History, 1 July - 31 December 1959, pp. 37, 265; 6555th Test Wing (Development) History, 21 December 1959 - 31 March 1960, "Organization" and "Mission."

32. 6555th Test Wing (Development) History, 21 December 1959 - 31 March 1960, "Personnel," DWSE Historical Section ("Function" and "Strength Resume") and DWTS Historical Section ("Introduction" and "Contract"); 6555th Test Wing (Development) History, 1 April - 30 June 1960, DWSM Historical Section; 6555th Test Wing (Development) History, 1 July - 31 December 1960, DWSE Historical Section.

33. 6555th Test Wing (Development) History, 21 December 1959 - 31 March 1960, DWOI Historical Section, p.6, DWTI Historical Section ("Introduction") and appendix ("Concept of Operations"), pp. 1, 2; 6555th Test Wing (Development) History, 1 April - 30 June 1960, DWO Historical Section.

34. 6555th Test Wing (Development) History, 21 December 1959 - 31 March 1960, DWTC Historical Section, ("Physical Facilities"); Marven R. Whipple, "Index of Missile Launchings by Missile Program, July 1960 - June 1961," 10 October 1961, p. 11-7.

35. 6555th Test Wing (Development) History, 21 December 1959 - 31 March 1960, DWTB Historical Section ("Physical Facilities"), DWTI Historical Section, ("Physical Facilities") and DWTM Historical Section ("Advanced Guidance Study Facilities"); 6555th Test Wing (Development) History, 1 July - 31 December 1960, DWTM Historical Section ("Physical

Facilities").

36. 6555th Test Wing (Development) History, 21 December 1959 - 31 March 1960, DWOC Historical Section, p. 3; 6555th Test Wing (Development) History, 1 April - 30 June 1960, DWOC Historical Section ("Missile Test Activities"); Whipple, "Index, July 1950 - June 1960," p. 11-5; Whipple, "Index, July 1960 - June 1961," pp. 11-5, 11-7, 11-8; Marven R. Whipple, "Atlantic Missile Range Index of Missile Launchings, July 1961 - June 1962," 26 October 1962, pp. 2, 3; 6555th Test Wing (Development) History, 1 January - 30 June 1961, DWTC Historical Section ("Introduction," "Personnel" and "Activities") and DWZC Historical Section ("Organization and Mission"); 6555th Aerospace Test Wing History, 1 July - 31 December 1961, DWTC Historical Section ("Introduction" and "Missile Test Activities"); Stanley,"<u>An Air Force Command for R&D,</u>" p. 41; Headquarters Space Division History Office, "Space and Missile Systems Organization: A Chronology 1954 - 1979," p. 4; Special Order G-174, HQ AFSC, 27 May 1963.

37. 6555th Aerospace Test Wing History, 1 January - 30 June 1962, DWTC Historical Section ("Personnel" and "Missile Test Activities"); Whipple, "Index, July 1961 - June 1962," p. 4; Marven R. Whipple, "Atlantic Missile Range Index of Missile Launchings, July 1962 - June 1963," undated, pp. 2, 3; 6555th Aerospace Test Wing History, 1 July - 31 December 1962, DWTC Historical Section ("Introduction," "Training" and "Test Activities").

38. Marven R. Whipple, "Atlantic Missile Range/Eastern Test Range Index of Missile Launchings," July 1963 - June 1964," undated, p. 2; Whipple, "Index, July 1962 - June 1963," p. 3; 6555th Aerospace Test Wing History, 1 January - 30 June 1963, DWTC Historical Section ("Introduction," "Training" and "Test Activities"); 6555th Aerospace Test Wing History, 1 July - 31 December 1963, DWTC Historical Section ("Training"); 6555th Aerospace Test Wing History, 1 January - 30 June 1964, DWTC Historical Section ("Training" and "Test Activities"); 6555th Aerospace Test Wing History, 1 July - 31 December 1964, "Organization" and DWX Historical Section ("Personnel" and "Missile Test Activities").

39. 6555th Test Wing (Development) History, 21 December 1959 - 31 March 1960, DWOB Historical Section; 6555th Test Wing (Development) History, 1 April - 30 June 1960, DWOB Historical Section ("Introduction" and "Major Problems"); 6555th Test Wing (Development) History, 1 July - 31 December 1960, DWOB Historical Section ("Major Problems" and "Program Activities") and DWTB Historical Section ("Program Activities" and "Key Personnel and Positions"); 6555th Test Wing (Development), 1 January - 30 June 1961, DWTB Historical Section ("Training" and "Program Activities"); Whipple, "Index, July 1960 - June 1961," pp. 12-4 and 12-5.

40. 6555th Test Wing (Development) History, 1 January - 30 June 1961, DWTB Historical Section ("Program Activities"); 6555 Aerospace Test Wing History, 1 July - 31 December 1961, DWTB Historical Section ("Training" and "Program Activities"); 6555th Aerospace Test Wing History, 1 January - 30 June 1962, DWTB Historical Section ("Program Activities") and DWTZ Historical

Section ("Physical Facilities"); Whipple, "Index, July 1961 - June 1962," p. 39.

41. 6555th Aerospace Test Wing History, 1 January - 30 June 1962, DWTB Historical Section ("Training" and "Program Activities"); Whipple, "Index, July 1961 - June 1962," p. 40; Whipple, "Index, July 1962 - June 1963," p. 38; 6555th Aerospace Test Wing History, 1 July - 31 December 1962, DWTB Historical Section ("Organization Structure," "Training" and "Program Activities"); 6555th Aerospace Test Wing History, 1 January - 30 June 1963, DWTB Historical Section ("Training" and "Program Activities").

42. 6555th Aerospace Test Wing History, 1 January - 30 June 1964, "Organization" and DWTB Historical Section ("Introduction," "Organizational Structure," "Personnel" and "Physical Facilities"); Whipple, "Index, July 1963 - June 1964," pp. 31, 32.

43. Neufeld, Ballistic Missiles, p. 182; 6555th Test Wing (Development) History, 21 December 1959 - 31 March 1960, DWTM Historical Section ("Introduction" and "Personnel"); 6555th Test Wing (Development) History, 1 April - 30 June 1960, DWTM Historical Section ("Missile Test Activities"); Article, "MINUTEMAN at a glance," Armed Forces Journal, 1 March 1971; 6555th Test Wing (Development) History, DWTM Section ("Missile Test Activities").

44. 6555th Test Wing (Development) History, 1 July - 31 December 1960, DWOM Historical Section ("Introduction," "Personnel" and "Problem Areas"); 6555th Test Wing (Development), 1 January - 30 June 1961, DWTM Historical Section ("Organization," "Personnel" and "Missile Test Activities"); 6555th Aerospace Test Wing History, 1 July - 31 December 1961, DWTM Historical Section ("Personnel" and "Training").

45. Whipple, "Index, July 1960 - June 1961," p. 28-2; Whipple, "Index, July 1961 - June 1962," pp. 20, 21; Whipple, "Index, July 1962 - June 1963," pp. 15 and 16; 6555th Aerospace Test Wing History, 1 January - 30 June 1962, DWTM Historical Section ("Missile Test Activities").

46. 6555th Aerospace Test Wing History, 1 January - 30 June 1962, DWTM Historical Section ("Missile Test Activities"); 6555th Aerospace Test Wing History, 1 July - 31 December 1962, DWTM Historical Section ("Organizational Structure," "Personnel" and "Missile Test Activities"); 6555th Aerospace Test Wing History, 1 July - 31 December 1963, DWTM Historical Section ("Personnel").

47. 6555th Aerospace Test Wing History, 1 January - 30 June 1963, DWTM Historical Section ("Strength Resume" and "Missile Test Activities").

48. Headquarters SAMTEC Office of History, "SAMTEC Master Chronology," o/a 1 June 1970, p. 3. Neufeld, Ballistic Missiles, pp. 237, 238; Headquarters SAC History Office, "SAC Missile Chronology, 1939 - 1988," pp. 37, 38, 41, 42, 48.

49. Whipple, "Index, 1 July 1963 - 30 June 1964," pp. 15-17; Marven R. Whipple, "Eastern Test Range Index of Missile Launchings, July 1964 - June 1965," undated, p. 15; 6555th Aerospace Test Wing History, 1 July - 31 December 1963, DWTM Historical Section ("Missile Test Activities"); 6555th Aerospace Test Wing History, 1 January - 30 June 1964, DWTM Historical Section ("Missile Test Activities"); 6555th Aerospace Test Wing History, 1 July - 31 December 1964, DWQ Historical Section ("Physical Facilities" and "Missile Test Activities"); 6555 Aerospace Test Wing History, 1 January - 30 June 1965, DWQ Historical Section ("Key Personnel," "Strength Resume" and "Test Activities"); 6555th Aerospace Test Wing History, 1 July - 31 December 1965, DWQ Historical Section ("Key Personnel," "Strength Resume" and "Test Activities"); 6555th Aerospace Test Wing History, 1 January - 30 June 1966, DWQ Historical Section ("Key Personnel" and "Test Activities"); 6555th Aerospace Test Wing History, 1 July - 31 December 1966, DWQ Historical Section ("Test Activities"); 6555th Aerospace Test Wing History, 1 January - 30 June 1967, DWQ Historical Section ("Key Personnel" and "Test Activities"); 6555th Aerospace Test Wing History, 1 July - 31 December 1967, DWQ Historical Section ("Test Activities"); 6555th Aerospace Test Wing History, 1 January - 30 June 1968, DWQ Historical Section ("Strength Resume" and "Missiles Launched").

50. 6555th Aerospace Test Wing, "Briefing Presented to AFSC IG Team (on) 4 November 1968," 5 November 1968; 6555th Aerospace Test Wing History, 1 January - 30 June 1968, DWQ Historical Section ("Strength Resume"); 6555th Aerospace Test Wing History, 1 July - 31 December 1968, DWQ Historical Section ("Strength Resume"); 6555th Aerospace Test Wing History, 1 July - 31 December 1969, DWQ Historical Section ("Strength Resume"); 6555th Aerospace Test Wing History, 1 July - 31 December 1970, DWQ Historical Section ("Strength Resume").

51. 6555th Aerospace Test Wing, "Briefing Presented to AFSC IG Team," 5 November 1968; AFSC Management Engineering Team Detachment 22, "6555th Aerospace Test Wing Organizational Chartbook," 15 October 1968, pp. 9, 10.

52. 6555th Aerospace Test Wing History, 1 July - 31 December 1968, DWQ Historical Section ("Personnel" and "Missile Operations"); 6555th Aerospace Test Wing History, 1 July - 31 December 1969, DWQ Historical Section ("Missile Operations"); Marven R. Whipple, "Eastern Test Range Index of Missile Launchings, July 1968 - June 1969," undated, p. 10; Marven R. Whipple, "Eastern Test Range Index of Missile Launchings, July 1969 - June 1970, undated, pp. 10, 11; Marven R. Whipple, "Eastern Test Range Index of Missile Launchings, July 1970 - June 1971," undated, p. 9; 6555th Aerospace Test Group History, 1 January - 30 June 1970, "Organization and Mission" and MINUTEMAN Test Division Historical Section ("Organization Structure," "Strength Resume" and "Missile Operations"); 6555th Aerospace Test Group History, 1 July - 31 December 1970, "Organization and Mission" and MINUTEMAN Test Division Historical Section ("Strength Resume" and "Missile Operations"); 6555th Aerospace Test Group History, 1 January - 30 June 1971, "Organization and Mission"; Space Systems Division History Office, "Space and Missile Systems Organization Chronology," pp. 4, 5, 12, 13.

The 6555th, Chapter IV, Section 1

Taking the High Ground: The 6555th's Role in Space through 1970

U. S. Military Space Efforts Through 1960

As we indicated in Chapter I, the Air Force's interest in artificial satellites -- and hence, space operations -- was sparked by discussions with the Navy shortly after the end of World War II. At Major General Curtis E. LeMay's request, the Douglas Aircraft Company's RAND group provided the Pentagon with a 321-page study in May 1946 on the feasibility of satellites for military reconnaissance, weather surveillance, communications and missile navigation. RAND considered the artificial satellite feasible, and the group predicted that the satellite would yield benefits in gravitational research, astronomy and bioastronautics as well as purely military operations. Though the Research and Development Board blocked the development of artificial satellites initially, RAND's research into the satellite's military usefulness continued from 1947 into the early 1950s. By November 1951, the Air Force had asked the Atomic Energy Commission to investigate the possibility of using small nuclear reactors as power sources for satellites, and the preliminary results of that investigation were favorable. The Radio Corporation of America (RCA) signed a contract with the RAND group in June 1952 to study optical systems, recording devices and imagery presentation techniques that might be used on reconnaissance satellites in the future. In July 1953, North American Aviation signed a contract with Wright Field's Communication and Navigation Laboratory to study a pre-orbital guidance system for satellites.[1]

The work up to that time had been concentrated on technological studies and analyses, but the Air Force's Air Research and Development Command redirected the satellite effort toward actual demonstrations of the satellite's major components as part of the Weapon System 117L program in the mid-1950s. Given the Air Force's commitment to the ATLAS program, the ATLAS missile was proposed as a suitable booster for the satellite, but the satellite program was kept on a separate track from the ATLAS so as not to delay the scheduled delivery of the ATLAS as a ballistic missile weapon system. Unfortunately, the Weapon System 117L program was funded in 1956 at only 10 percent of the level needed to meet its requirements in 1957 (e.g., $3 million versus $39.1 million). Thus, despite a sound technical foundation, the Air Force satellite effort suffered from inadequate funding through 1957.[2]

The Soviets' successful launch of Sputnik I on 4 October 1957 came as a shock to the American public, but the military implications of that capability came into even sharper focus as much heavier payloads were orbited from the Soviet Union in the months and years that followed. Galvanized into action by the Soviet Union's achievements, the U.S. Department of Defense set high priorities on the development of military satellite systems. It also created the Advanced Research Projects Agency (ARPA) on 7 February

1958 to supervise all U.S. military space efforts. The Air Force drew up a manned military space system development plan in April 1958, and it also volunteered to carry out the U.S. man-in-space mission. Though much of the plan was incorporated in later manned space efforts (e.g., MERCURY, GEMINI and APOLLO), President Eisenhower rejected the Air Force's offer to lead the effort. Instead, he called on Congress to establish a civilian space agency, and the National Aeronautics and Space Act was passed by Congress in July 1958. Under Executive Order Number 10783, the National Aeronautics and Space Administration (NASA) became the controlling agency for non-military scientific space projects on 1 October 1958. The Navy's VANGUARD satellite project and ARPA's lunar probe project were transferred to NASA on October 1st, but ARPA retained its military satellites, high energy rocket upper stages and its military space exploration programs. The Advanced Research Projects Agency also expected to participate in the early stages of the manned space program along with NASA and the individual military branches.[3]

Under ARPA, the Weapon System 117L satellite program was divided into three R&D packages:

- 1) DISCOVERER (A THOR-boosted satellite system used to orbit biomedical payloads and engineering experiments.)
- 2) SENTRY (An ATLAS-boosted reconnaissance satellite program.)
- 3) MIDAS (The ATLAS-boosted MIssile Defense Alarm System which was designed to use infrared sensors to detect ballistic missile plumes and provide early warning in the event of a missile attack.)

During 1959, the Air Force supported all three programs, as well as ARPA's TRANSIT program, which used satellites to investigate the earth's cloud cover from orbit, and the COURIER program, which orbited experimental communications repeater systems.[4]

Since the Air Research and Development Command was destined to serve the Air Force and two non-Air Force clients in space (i.e., ARPA and NASA), effective coordination among the three agencies was crucial to the early success of the space mission. At first, the ARDC Commander assigned ARPA point-of-contact duties to his executive officer, but the scope and nature of the job suggested its placement elsewhere, and it was assigned to ARDC's Weapon Systems Analysis Division Chief, Lieutenant Colonel Ralph J. Hicks, on 25 August 1958. (Lieutenant Colonel Hicks also served as the ARDC point-of-contact for NASA, and he played an important role in coordinating discussions of mutual concern with ARPA and NASA.) After Lieutenant General Schriever assumed command of ARDC in April 1959, he elevated the point-of-contact position and changed its title to Special Assistant to the Commander for ARPA and NASA Affairs on 20 July 1959. As such, the Special Assistant's actions were to be treated as if they came from Lieutenant General Schriever himself. Space activities became increasing important to the ARDC mission, and the Command's effective management of space boosters and interagency space support requirements in the late 1950s and 1960 had much to do with Secretary McNamara's decision to give the Air Force responsibility for the Department of Defense's portion of the national space program on 6 March 1961. As a result of those management actions and decisions, the 6555th and its contractors were placed on the cutting edge of space operations at Cape Canaveral in the

1960s. The 6595th Aerospace Test Wing and its contractors were in a similar position at Vandenberg.[5]

Before the 6555th absorbed the Air Force Ballistic Missile Division's resources at the Cape in December 1959, most of the Air Force's participation in the Cape's space launch operations was managed by the WS-315A (THOR) Project Division under the Air Force Ballistic Missile Division's Assistant Commander for Missile Tests. The WS-315A Project Division was redesignated the Space Project Division on 16 November 1959, and it became the Space Projects Division under the 6555th Test Wing on 15 February 1960. Both actions were prompted by the demise of THOR ballistic missile testing at the Cape, not a fundamental change in the Division's space support operations. Though the Division consisted of less than a dozen officers and clerical personnel, it was charged with monitoring the Douglas Aircraft Company's booster preparations and launch operations, and it coordinated range support for space missions. The Division had jurisdiction over Complex 17 and three missile assembly buildings (e.g. hangars M, L and AA). Lieutenant Colonel Thomas W. Morgan served as the Division's chief from 2 April 1958 onward, and the Division supported a total of 10 Air Force-sponsored THOR-ABLE, THOR-ABLE I and THOR-ABLE II space launches from Pad 17A before the end of 1959. The Division also supported NASA's PIONEER I and II missions, which were launched by Douglas from Pad 17A on 11 October and 8 November 1958, and NASA's EXPLORER VI mission, which was launched by Douglas from Pad 17A on 7 August 1959. Under the 6555th Test Wing (Development), the Space Projects Division managed five THOR-ABLE-STAR missions for the Army, the Navy and ARPA in 1960. It also monitored Douglas' preparation and launch of two THOR-ABLE boosters for NASA's PIONEER V deep space mission to Venus in March 1960 and its TIROS I weather satellite mission in April 1960.[6]

COMPLEX 17 SHOWING PADS 17A AND 17B
1961

THOR-ABLE LAUNCH OF TIROS I SATELLITE FROM PAD 17A
1 April 1960

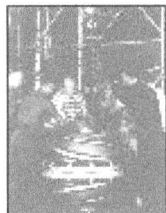
WEIGHTS AND CENTER OF GRAVITY TEST - TRANSIT SATELLITE
28 March 1960

THOR-ABLE-STAR LAUNCH OF TRANSIT 1B FROM PAD 17B
13 April 1960

THOR-ABLE-STAR PREPARED FOR ARMY TRANSIT 3A SATELLITE MISSION
30 November 1960

THOR-ABLE-STAR LAUNCH OF TRANSIT 3A
30 November 1960

The 6555th used several other divisions to handle other parts of its space support mission in 1959 and 1960. Though the Cape's role in the MIDAS satellite R&D project was relatively short-lived, the 6555th assigned a handful of its personnel to its MIDAS Project Division to monitor Lockheed Aircraft Corporation's activities at Complex 14 and Hangar E at the end of 1959. The TS 609A Project Division was activated under the Assistant Commander for Missile Tests in August 1959 to monitor Aeroneutronic's solid rocket operations on Complex 18, and the TS 609A Operations Division was added under the 6555th Test Wing in December 1959 to develop a blue suit capability to assemble, maintain, checkout and launch TS 609A launch vehicles. Under the 6555th's Directorate of Support, the Facilities Division monitored all new Air Force Ballistic Missile Division construction at the Cape, including Space Launch Complex 36, which was being built for the CENTAUR program in 1960 and 1961. In the spring of 1960, the Space Projects Division's responsibilities were broadened to include program planning for NASA's ATLAS/AGENA-B program at Cape Canaveral. Hangar E, which had been associated with the MIDAS project, was assigned to Lockheed Aircraft Corporation for NASA's RANGER project in the summer of 1960. The 6555th faced more responsibilities in 1961 as the space program continued to grow.[7]

The 6555th, Chapter IV, Section 2

Taking the High Ground: The 6555th's Role in Space through 1970

ATLAS, THOR, and BLUE SCOUT Space Operations

Following the establishment of Air Force Systems Command, the 6555th's Test Directorate and Operations Directorate were transformed into the Space Programs Office and the Ballistic Missiles Office on 17 April 1961. Under that reorganization, the old ATLAS Project Office's resources were divided roughly in half to create an ATLAS Booster Branch and an ATLAS Weapons Branch. The ATLAS Booster Branch was placed under the Space Programs Office. The old ATLAS Operations Division became the new ATLAS Weapons Branch's Operations Section, and the new ATLAS Weapons Branch was placed under the Ballistic Missiles Office. The Space Projects Division became the Space Projects Branch under the Space Programs Office on April 17th, and its THOR Booster Branch (created on 17 March 1961) was removed and set up as a separate branch under the Space Programs Office. The TS 609A Project Division and TS 609A Operations Division were combined to form the BLUE SCOUT Branch, which was also placed under the Space Programs Office. The TITAN, MINUTEMAN and MACE Operations Divisions became the TITAN, MINUTEMAN, and MACE Weapons Branches, and they were placed under the Ballistic Missiles Office. Though the Space Programs Office and the Ballistic Missiles Office were renamed on 25 September 1961, the separation of operations into 1) space projects and 2) ballistic missile test operations continued even after the 6555th's branches were upgraded to divisions in the summer of 1962. The 6555th's Technical Support Office managed the Wing's budget and supplies and monitored its facilities and technical requirements.[8]

MIDAS NOSE SECTION ON TRANSPORTER AT PAD 14

14 December 1959

MIDAS NOSE SECTION IN PLACE ON GANTRY AT PAD 14

14 December 1959

Under Lieutenant Colonel Harold A. Myers, the 6555th's Space Projects Branch focused its attention on satellites and spacecraft being prepared by contractors at Missile Assembly Hangar AA for the

TRANSIT, ANNA, RANGER, SAINT and VELA HOTEL projects. Space boosters were monitored by the Space Program Office's other three branches until 25 September 1961, when the THOR Boosters Branch and the Space Projects Branch were recombined to form the THOR/TITAN Space Branch. Thereafter, the Deputy for Space Systems accomplished his space mission through the THOR/TITAN Space Branch, the ATLAS Space Branch and the BLUE SCOUT Branch.[9]

ATLAS "D" LAUNCH OF MIDAS MISSION

26 February 1960

By any standards, 1961 proved to be a very busy year for the 6555th and its space launch contractors. Following its first two unmanned MERCURY capsule launches for NASA in September 1959 and July 1960, Convair launched ATLAS boosters on three successful (and one unsuccessful) MERCURY flights from Complex 14 in 1961. The Douglas Aircraft Company launched three TRANSIT navigation satellite missions from Pad 17B for the U.S. Navy, and it provided booster support for two EXPLORER missions and one TIROS mission that were launched from Pad 17A in 1961. Aeroneutronic and the BLUE SCOUT Branch's Operations Section launched a total of six space vehicles from pads 18A and 18B in 1961. NASA Associate Administrator Robert C. Seamans, Jr. signed a joint NASA/ARDC agreement on 30 January 1961 concerning the Air Force's participation in the AGENA B Launch Vehicle Program, and the 6555th's participation in the CENTAUR program was settled with NASA under a joint memorandum of agreement in April 1961. Both agreements were enormously important to the 6555th's role in the U.S. space program, and they set the tone for the Wing's space operations in the 1960s.[10]

FIRST ATLAS MERCURY CAPSULE LAUNCH FROM PAD 14

9 September 1959

DR. KURT DEBUS

Under the AGENA B Launch Vehicle Program agreement, NASA confirmed that it was pursuing the AGENA program through "established USAF Satellite System channels" to take advantage of Air Force capabilities and procedures. In effect, while NASA's Office of Launch Vehicle Programs retained overall management authority for AGENA, its Assistant Project Director for AGENA coordinated AGENA B requirements through the Air Force Ballistic Missile Division and its contractors. The Air Force Ballistic Missile Division procured the ATLAS boosters required by the program, and it provided operational, administrative and technical support for those launch vehicles. NASA's Jet Propulsion Laboratory and the Goddard Space Flight Center provided the spacecraft. The Launch Operations Directorate's Test Support Office acted as NASA's formal point of contact for all agencies involved in the AGENA B program on the Eastern Test Range, but the 6555th was responsible for supervising the Air Force contractors who provided the boosters for the AGENA B. While many tests were observed jointly by NASA and Air Force representatives, NASA was responsible for the spacecraft, Lockheed was responsible for the AGENA B, Convair was responsible for the ATLAS booster, and the 6555th was responsible for the readiness of the entire launch vehicle. Ultimately, NASA's Operations and Test Director had overall responsibility for the countdown, but he received direct inputs from the 6555th's Test Controller concerning the vehicle's status on launch day. The 6555th and the Air Force's contractors were thus very close to the center of the entire operation.[11]

Because the CENTAUR was conceived by ARDC and sponsored by the Air Force before ARPA adopted it as a DOD-wide project, Air Force Systems Command took a keen interest in the CENTAUR's development as an upper stage rocket. The project was eventually transferred to NASA, but the Air Force continued to follow the CENTAUR' development closely, and it loaned some officers to NASA to help manage the project. The Air Force also earmarked some of its test facilities on Cape Canaveral for the ATLAS/CENTAUR effort. While the CENTAUR agreement signed in April 1961 acknowledged NASA's authority to exercise launch responsibility for all 10 CENTAUR R&D test vehicles and all CENTAURs used on NASA's operational missions, it also confirmed that the 6555th Test Wing would exercise similar launch responsibilities for CENTAURs used on operational DOD missions. The 6555th was also allowed to assign Air Force supervisors to Convair's processing teams while they were working on ATLAS "D" boosters for the ATLAS/CENTAUR R&D test flights. In instances where NASA's Launch Operations Directorate wanted procedures added to Convair's ATLAS checklists, the 6555th integrated those items. NASA also agreed to coordinate CENTAUR test documentation with the 6555th. To avoid duplication of effort, NASA and the Air Force agreed to share "a large number of facilities" (e. g., Complex 36 and hangars H, J and K) for the CENTAUR, AGENA B and MERCURY efforts. Since NASA planned to use the CENTAUR's facilities first, the Air Force secured a promise from NASA to coordinate its CENTAUR facility and equipment modifications with the 6555th before the changes were made. The 6555th agreed to make an officer available as a consultant to NASA's Launch Director during ATLAS/CENTAUR launch operations.[12]

As Chief of the ATLAS Booster Branch, Lieutenant Colonel Robert R. Hull handled most of the 6555th's AGENA B, CENTAUR and Project MERCURY responsibilities with a staff consisting of an ATLAS Systems Section Chief, two civilian secretaries and six project officers. Within a few months, however, it became clear that the 6555th's space activities needed to be reorganized: the ATLAS Booster Branch was renamed the ATLAS Space Branch on 15 August 1961, and Lieutenant Colonel Hull picked up five additional officers and two secretaries from the Space Projects Branch as part of that branch's transformation into the THOR/TITAN Space Branch in late September 1961. Management of Complex 36A was transferred from the 6555th to NASA in the fall of 1961, but the Air Force retained ownership of the facility, and Colonel Wignall reaffirmed the 6555th's right to approve any modifications to Complex 36 before NASA carried them out. Though the ATLAS Space Branch acted only as a technical consultant for NASA's CENTAUR development program, it supported NASA's ATLAS booster requirements in accordance with the Seamans/Schriever agreement. The Branch also retained jurisdiction over military missions involving the ATLAS "D" and AGENA B upper stage as space boosters.[13]

In 1962, Air Force contractors and the ATLAS Space Branch supported three RANGER and two MARINER missions from Complex 12, and they supported the first three manned orbital ATLAS/MERCURY missions, which were launched from Complex 14. All those NASA missions were launched by contractors, but the Air Force implemented plans in the last half of 1962 to establish an ATLAS/AGENA "blue suit" launch capability. In September, the 6555th started negotiations with Air Training Command and the Air Force's contractors to establish an ATLAS/AGENA military launch team. One hundred and thirty-six enlisted positions were approved for the effort, and six airmen were assigned to the SLV-III Division (formerly the ATLAS Space Branch) by the end of 1962. Though the SLV-III Division gained 151 additional men (including 11 veterans from the ATLAS Weapons Division) over the next six months, the Air Force suddenly dropped the plan and terminated the ATLAS/AGENA blue suit program on 1 January 1964. The SLV-III Division's strength was reduced to approximately two dozen officers, airmen and civilians by the middle of 1964, and it remained near that level for the balance of the 1960s. The Division's MERCURY support mission ended following the last MERCURY flight in May 1963, but the unit still supported DOD operations on Complex 13. It picked up ATLAS/AGENA Target Vehicle operations for Project GEMINI shortly thereafter.[14]

COMPLEX 36 BLOCKHOUSE AND CABLEWAY
January 1960

Under Major (and later, Lieutenant Colonel) LeDewey E. Allen, Jr., the SLV-III Division supported three DOD satellite missions from Complex 13 in October 1963, July 1964 and July 1965. The Division also supported seven ATLAS/AGENA target vehicle missions launched from Complex 14 between 25 October 1965 and 12 November 1966. The Division's radio guidance station provided services for many NASA missions during the 1960s, and the Guidance Branch was the only SLV-III element involved in launch operations in 1967. Under Lieutenant Colonel Earl B. Essing, the SLV-III Division rallied in 1968 to support four classified DOD launches from Complex 13 between 6 August 1968 and 31 August 1970. The SLV-III Division became the ATLAS Systems Division in 1970, but this redesignation had little, if any, affect on the mission. Lieutenant Colonel Essing's people continued their work as 300 contractor personnel prepared for more DOD missions from Complex 13 in the 1970s.[15]

COMPLEX 36

January 1961

Though a blue suit launch capability for ATLAS space vehicles was never developed at Cape Canaveral, the 6555th retained a limited military launch capability for "guided" BLUE SCOUT and "unguided" BLUE SCOUT JUNIOR space vehicles during the first half of the 1960s. Airmen assisted on several BLUE SCOUT launches in 1961, and the first all-military BLUE SCOUT processing operation was completed in April 1962, shortly after Lieutenant Colonel Millard E. Griffith succeeded Lieutenant Colonel Jesse G. Henry as BLUE SCOUT Branch Chief. The Branch had more than a hundred people assigned to its activities before the BLUE SCOUT program was deleted in favor of the BLUE SCOUT JUNIOR program in July 1962. (The unit managed to hold on to 7 officers, 68 airmen and 7 civilians for the BLUE SCOUT JUNIOR program after the Branch became the SLV-IV Division on 1 August 1962.) Under Lieutenant Colonel John B. Adams, the Division's Combined Systems Test Building and Assembly and Checkout Building were accepted on 13 March 1963, and BLUE SCOUT JUNIOR modifications to Pad 18A were completed in June 1963. Factory and on-the-job training was underway by the spring of 1963, and the first all-military launch of a BLUE SCOUT JUNIOR was completed successfully on 30 July 1963. Under Lieutenant Colonel Warren L. Foss, the Division launched another BLUE SCOUT JUNIOR on 13 March 1964, but a fourth stage failure marred the flight, and very little data was obtained on the mission. Late deliveries and component flaws handicapped the Division's best efforts, but the program's last five space vehicles were launched (with mixed results) between 28 January and 10 June 1965. Its mission completed, the SLV-IB Division disbanded in the last half of 1965, and its personnel were transferred to other agencies under the Wing.[16]

CENTAUR LAUNCH PADS 36A AND 36B

1964

ATLAS-AGENA B LAUNCH OF RANGER IV FROM PAD 12

23 April 1962

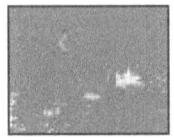
ATLAS-AGENA LAUNCH OF DOD SATELLITE FROM PAD 13

20 July 1965

ATLAS-AGENA LAUNCH OF GEMINI TARGET VEHICLE FROM PAD 14

16 March 1966

COMPLEX 14 ATLAS SPACE BOOSTER CREDITS

Through September 1966

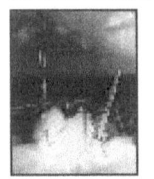
FIRST LAUNCH OF BLUE SCOUT JR. FROM PAD 18A

21 September 1960

BLUE SCOUT ON PAD 18A

19 May 1961

BLUE SCOUT LAUNCH

9 May 1961

BLUE SCOUT JR. BEING PREPARED FOR LAUNCH FROM PAD 18A

30 July 1963

As we noted earlier, the THOR Booster Branch was dissolved on 25 September 1961, and its resources were absorbed by the THOR/TITAN Space Branch. In addition to satellite missions, the new 15-member THOR/TITAN Space Branch became involved in the DYNA SOAR, GEMINI, and TITAN III programs. The Branch was reorganized under Lieutenant Colonel Jesse G. Henry, and complexes 19 and 20 were added to the Branch's holdings during the first half of 1962 for TITAN II space operations. On

10 September 1962, the Wing established the SLV-V Division to handle the TITAN III program separately, and it transferred TITAN III personnel from the THOR/TITAN Branch to the new division before renaming it the SLV-V/X-20 Division on 1 October 1962. The THOR/TITAN Branch became the SLV II/IV Division under Lieutenant Colonel Robert R. Hull on 1 October 1962, but it was split up to form two new divisions -- the SLV II Division (for THOR) and the GEMINI Launch Vehicle Division (for TITAN II) -- on 20 May 1963. By the end of June 1963, Major Carl B. Ausfahl was in charge of the GEMINI Launch Vehicle Division, a staff of eight people, Complex 19 and parts of hangars T, U and G. During the same period, Major Robert B. Gallman was in charge of ten officers, two airmen and seven civilians at the SLV-V/X-20 Division's offices in Hangar G. Rounding out the reorganization, Major Richard W. Marshall was in charge of Complex 17, other THOR facilities, four officers and three civilians at the SLV-II Division at the end of June 1963. (Marshall was promoted to Lieutenant Colonel shortly thereafter, and the SLV-II Division's strength grew to nine officers, three airmen and five civilians by the end of the year.) In the midst of the 6555th's organizational changes, Air Force contractors launched three Navy satellite flights and supported eleven NASA space missions from Complex 17 in 1962. Air Force contractors also provided space booster support for seven NASA THOR/DELTA missions, and they launched the first THOR/ASSET experimental reentry vehicle on a successful mission in 1963.[17]

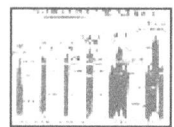 *TITAN FAMILY OF MISSILES AND SPACE BOOSTERS*

Though Complex 17 supported seven other NASA missions in 1964 and 1965, the two-part ASSET (Aerothermodynamic/Elastic Structural Systems Environmental Tests) program quickly became the SLV-II Division's principal interest after the first ASSET launch on 18 September 1963. Under one part of the ASSET flight test program at the Cape, the second, third, and sixth hypervelocity vehicles were launched from Pad 17B on 24 March 1964, 22 July 1964 and 23 February 1965. Those flights were designed to gather data on the ability of materials and structures to handle the pressures and temperatures of atmospheric reentry. Though the flight on March 24th failed to meet its test objectives due to a malfunction in the THOR's upper stage, the other two flights were successful, and the vehicle launched on July 22nd was recovered. Under the other part of the ASSET flight testing, two non-recoverable delta wing glide vehicles were launched from Pad 17B on 27 October and 8 December 1964. Both missions were designed to obtain data on "panel flutter" under high heating conditions and information of the vehicles' "unsteady aerodynamics" over a broad range of hypersonic speeds. Both flights were successful, and the final ASSET flight on 23 February 1965 completed the ASSET program.[18]

THOR-ASSET LAUNCH VEHICLE ON PAD 17B

28 August 1963

The Air Force had no further use for THOR facilities at Cape Canaveral after the ASSET program was completed, so the Space Systems Division directed the 6555th to turn over its SLV-II facilities to NASA for the civilian agency's THOR/DELTA program. In accordance with Air Force Eastern Test Range procedures, the 6555th returned the facilities to the Range in April 1965, and the Air Force Eastern Test Range transferred them to NASA's Kennedy Space Center in May 1965. The 6555th retained jurisdiction over other facilities that supported several programs in addition to THOR (e.g., Building 1381, the Hangar AA storage area and the Western Electric Guidance Facility), and it reassigned most of the SLV-II Division's personnel to the Eastern Test Range or to other duties in the Wing. Two of the Division's officers were reassigned to other Air Force bases, and Lieutenant Colonel Marshall went on retirement leave in May 1965. Once the administrative details of those actions were completed, the SLV-II Division was phased out in June 1965.[19]

FINAL ASSET PAYLOAD INSTALLED ON THOR BOOSTER AT PAD 17B

23 February 1965

The 6555th, Chapter IV, Section 3

Taking the High Ground: The 6555th's Role in Space through 1970

The TITAN II / GEMINI Program

The 6555th's TITAN II/GEMINI Division lasted considerably longer than its SLV-II Division. Though no TITAN II/GEMINI launch operations were supported in 1963, Air Force contractors completed an 18-month-long remodeling project on Complex 19 for Project GEMINI in August. The GEMINI Launch Vehicle Division monitored that effort, and it supervised the Martin Company's checkout of the first GEMINI launch vehicle, which arrived from Martin's Baltimore facility on 26 October 1963. Under Lieutenant Colonel (and later, Colonel) John G. Albert, the Division exercised technical test control over the TITAN II/GEMINI launch vehicle, but the Martin Company launched the booster. Martin launched the first unmanned TITAN II/GEMINI mission from Complex 19 on 8 April 1964, and the flight succeeded in placing an unmanned 7,000-pound GEMINI capsule into low earth orbit on that date. Though the second unmanned TITAN II/GEMINI launch vehicle had to be dismantled to protect it from two hurricanes in August and September 1964, it was launched successfully on 19 January 1965. The first manned GEMINI mission was launched from Complex 19 on 23 March 1965, and it met all of its test objectives. (Astronauts Virgil I. Grissom and John W. Young were recovered with their capsule in the primary recovery area after three orbits on March 23rd.) Nine more pairs of astronauts were boosted into orbit aboard TITAN II/GEMINI launch vehicles in 1965 and 1966, and seven ATLAS/AGENA target vehicles were launched from Complex 14 in support of six GEMINI missions.[20]

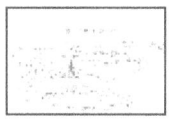
GEMINI/TITAN II COMPLEX 19

Following the last highly successful TITAN II/GEMINI flight in November 1966, the GEMINI Launch Vehicle Division completed its mission and began transferring personnel to other Air Force bases or to other agencies under the 6555th Aerospace Test Wing. Colonel Albert became the 6555th's Executive Officer toward the end of January 1967, and the Wing was awarded the Theodore von Karman trophy for its successful support of Project GEMINI. Though GEMINI had been a NASA project, DOD experiments were conducted on GEMINI flights, and NASA and the Air Force both derived benefits from the astronauts' activities in space. Like Project MERCURY, Project GEMINI demonstrated that space launch vehicles could be operated with a high degree of safety, reliability and precision. GEMINI also provided valuable experience in rendezvous and docking techniques -- skills that would be vital to astronauts returning from expeditions to the surface of the moon. As overall manager for Project

GEMINI, NASA was understandably proud of its role in the highly successful effort, but the Air Force and its contractors planned, built and launched all the TITAN II space boosters associated with Project GEMINI. Unfortunately, this distinction was often overlooked by the public and the news media on launch day.[21]

TITAN II-GEMINI LAUNCH CREW LED BY LT. COLONEL ALBERT

March 1964

*COMPLEX 19
BLOCKHOUSE
August 1963*

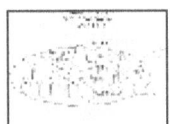

GEMINI/TITAN II FIRST FLOOR BLOCKHOUSE COMPLEX 19

TEST CONDUCTOR CONSOLE BLOCKHOUSE 19

November 1961

*MISSION TABLE SHOWING MANNED MISSION CREDITS
THROUGH SEPTEMBER 1966*

2ND UNMANNED GEMINI CAPSULE LAUNCH

19 January 1965

*NASA GROUP ACHIEVEMENT AWARD PRESENTED TO GEMINI LAUNCH
VEHICLE DIVISION*

Left To Right: Mr. Westmoreland (Pan Am/RCA), Mr. Peer (General Electric), Mr. Barnes (Aerojet), Mr. Cauldwell (Burroughs), Major Henry (6555th), Mr. Cary (Martin), Colonel Ledford (6555th Commander), Colonel Warner (Patrick Test Site Office), Mr. Wiegard (Aerospace) And Lt. Colonel Albert (Gemini Launch Vehicle Division Chief).

The 6555th, Chapter IV, Section 4

Taking the High Ground: The 6555th's Role in Space through 1970

The TITAN III Program

The TITAN III program, however, was a different matter. NASA's plans for the SATURN program were already underway in 1961, and the agency saw no need for another heavyweight space booster for low earth orbit, geosynchronous orbit or deep space missions. Consequently, NASA resisted the Air Force's first attempts to secure funding for the TITAN III initiative, and the Air Force had to work long and hard to prepare its case for the TITAN III. Inter-agency discussions did not reach a turning point until November 1961, when Dr. N.E. Golovin's Large Launch Vehicle Planning Group concluded that the TITAN III was the best booster for DOD missions after 1963. NASA still did not concede any need for the TITAN III until the Golovin Group modified its recommendation on December 5th to conclude that the TITAN III was the best choice for DOD missions after 1965. (Even at that juncture, NASA only considered the TITAN III a gap-filler between the ATLAS/CENTAUR and the much larger SATURN space booster.) Following that concession, initial funding for the TITAN III contractual effort was granted on 11 December 1961, and Space Systems Division's new 624A Systems Program Office began managing the TITAN III program four days later.[22]

As engineering efforts proceeded elsewhere, NASA and the Department of Defense signed an agreement in January 1963 which acknowledged the Air Force's jurisdiction over all TITAN III construction at the north end of Cape Canaveral. Though TITAN Complex 41 extended across the Cape Canaveral boundary into NASA's territory on Merritt Island, all property within Complex 41's security fence and along the access road to the site was considered part of the Air Force's Titan III program. Put simply, NASA had jurisdiction over the Merritt Island Launch Area, the SATURN program and SATURN facilities on Merritt Island and Cape Canaveral. The Air Force had jurisdiction over Cape Canaveral, the TITAN III program and all TITAN III facilities, including Complex 41. Though the Air Force Eastern Test Range and its contractors continued to provide range support for all of NASA's launch vehicle programs on Merritt Island and Cape Canaveral, the SATURN and TITAN III programs were pursued as distinctly separate NASA and Air Force launch efforts.[23]

The Cape's TITAN IIIC construction program began in earnest on 24 November 1962 after a $4.6 million contract was awarded to the Atlantic Gulf and Pacific Company to prepare sites for launch complexes 40 and 41. Though the timetable for completion of the pads remained "soft" for several months, both complexes had to be connected to other facilities in the TITAN IIIC Integrate-Transfer-Launch (ITL) system via railroad lines across a shallow area in the Banana River. Since much of the land selected for other ITL facilities was also covered by shallow water, most of the area had to be built up, and dredging operations were underway by February 1963 to move 6.5 million cubic yards of landfill from the Banana River to the ITL sites. The contract for the TITAN IIIC launch complexes was awarded to C. H. Leavell and Peter Kiewit & Sons on 13 June 1963, and it was completed in 1965 for approximately $17 million. A $479,500 contract for the TITAN IIIC railroad (i.e. an extension of the NASA railroad network) was awarded to B. B. McCormick, Inc. on 30 July 1963, and it was completed in 1964. Most of the other ITL facilities were grouped under a $26.8 million contract awarded to the firm of Paul Hardeman and Morrison-Knudsen on 30 July 1963. That contract was completed on 16 April 1965.[24]

DREDGES IN OPERATION AT CAPE ITL AREA

VERTICAL INTEGRATION BUILDING AREA

1963

AERIAL VIEW OF ITL AREA

29 April 1964

SMAB AND VIB CONSTRUCTION

22 April 1964

COMPLEX 40 PAD AREA LOOKING EAST

2 January 1964

TITAN III LOCOMOTIVES

16 November 1964

AERIAL VIEW OF TITAN IIIC AREA

November 1964

COMPLEX 20 TITAN IIIA PAD CREW

October 1964

FIRST LAUNCH OF TITAN IIIA FROM PAD 20

1 September 1964

To coordinate those construction efforts and prepare for the TITAN IIIC's debut at Cape Canaveral, the 6555th formed a task force on 15 January 1963. The task force was led by Major Robert B. Gallman initially, and it was composed of all the staff and operating units of Gallman's TITAN III/X-20 Division, three officers from the 6555th's Missile Test Facilities Office and the Aerospace Corporation's 624A Project Office. In addition to the ITL area construction, the task force monitored an $819,000 contract with Julian Evans and Associates to modify Complex 20 for TITAN IIIA operations. That contract was completed to the point of beneficial occupancy in September 1963, and Martin's sub-contractors completed equipment installation and ground systems testing on Complex 20 by the end of June 1964. Martin launched the first TITAN IIIA from Complex 20 on 1 September 1964, and three more TITAN IIIA flights were completed before the first TITAN IIIC was launched from Complex 40 on a successful mission on 18 June 1965. Following the fourth and final TITAN IIIA launch on 6 May 1965, Complex 20 was deactivated and returned to the Air Force Eastern Test Range in September 1965. Complex 41 was turned over to the TITAN III Division's Operations Branch for beneficial occupancy on 18 June 1965, and the facility was accepted by the Air Force in December 1965.[25]

TITAN IIIC AND GANTRY ON PAD 40

23 MAY 1965

LAUNCH OF 1ST TITAN IIIC FROM COMPLEX 40

18 June 1965

As ITL construction got underway in the summer of 1963, officers and men poured into the TITAN III/X-20 Division to oversee the work. Among the new arrivals was Lieutenant Colonel Marc M. Ducote, who replaced Major Gallman as Division Chief. Major Gallman continued to serve the Division as its Test Support Branch Chief, and Major E. J. D'Arcy arrived in October 1963 to serve as the Division's Flight Test Operations Branch Chief. The Division had 31 officers, 18 airmen and 14 civilians assigned to monitor the TITAN III program by the end of 1963, and the personnel roster grew to 39 officers, 31 airmen and 14 civilians by the middle of 1964. Following cancellation of the DYNA SOAR project in 1963, the Division was renamed the TITAN III Division, and it was reorganized into three branches in 1964 to provide more efficient supervision of the contractors' efforts. Under the new organization, Lieutenant Colonel Ducote continued as Division Chief, and Major D'Arcy ran the Systems Branch. Major Edwin E. Speaker served as Operations Branch Chief, and Major Ralph S. Davison managed the Test Support Branch. In the fall of 1964, four SAC officers were assigned to the Division to learn more about the TITAN III system before SAC played host to TITAN IIIB and TITAN IIID launches at Vandenberg Air Force Base. Rounding out the year in December, the TITAN III Division gained a new agency -- the Manned Orbiting Laboratory Branch -- with Major Joseph R. Henry as its chief. The MOL Branch became the Payloads Branch in the last half of 1965, and it assumed responsibility for all TITAN IIIC payloads.[26]

TITAN III OPERATIONS CONTROL CREW

14 October 1965

TITAN IIIC LAUNCH FROM PAD 40

15 October 1965

Following the first TITAN IIIC launch in June 1965, service contractors refurbished Complex 40 for the Cape's second TITAN IIIC flight. Though the second TITAN IIIC launch went well on 15 October 1965, the vehicle suffered a malfunction at the start of a transfer orbit, and it failed to complete its experimental mission. During the same period, the Martin Company and its sub-contractors were hard at work on Complex 41 to prepare that facility for its first TITAN IIIC launch in December. Complex 41 was accepted by the Air Force on 15 December 1965, and the first TITAN IIIC lifted off Pad 41 on December 21st. The flight met most of its test objectives, including the successful release of the LES-3 and LES-4 communications satellites and the OSCAR IV (amateur radio) satellite. Two more TITAN IIIC missions were launched from Complex 41 on 16 June and 26 August 1966. The first of those flights included the successful release of seven Initial Defense Communications Satellite Program (IDCSP) satellites and one gravity gradient satellite, but the second flight ended after the TITAN IIIC's payload fairing broke up approximately 79 seconds after launch. (Eight IDCSP satellites were destroyed in the mishap.) Another TITAN IIIC was launched from Complex 40 on 3 November 1966, and it boosted a modified GEMINI spacecraft and three secondary satellites into orbit during a largely successful experimental mission on that date.[27]

COMPLEX 41
26 August 1966

Since the Air Force intended to use Complex 40 for its Manned Orbiting Laboratory (MOL) flights, Complex 41 eventually supported all the TITAN IIIC missions launched from the Cape between the beginning of 1967 and the end of the decade. On the first of those missions, a TITAN IIIC boosted eight IDCSP satellites into orbit on 18 January 1967. Martin also launched two VELA nuclear detection satellites and three environmental research satellites aboard a TITAN IIIC on April 28th, and it launched three more IDCSP satellites, two other communications satellites and a Department of Defense Gravity Experiment (DODGE) on 1 July 1967. In 1968, a TITAN IIIC placed eight IDCSP satellites into orbit on June 13th, and another TITAN IIIC boosted the LES-6 communications satellite and three scientific satellites into various orbits on September 26th. The next TITAN IIIC mission placed a 1,600-pound Air Force communications satellite into orbit on February 9th, and another TITAN IIIC boosted two VELA satellites and three experimental satellites into orbit on 23 May 1969.[28]

TITAN IIIC GEMINI/MOL
LAUNCH
3 November 1966

TITAN IIIC IDCSP SATELLITE MISSION FROM PAD 41

18 January 1967

No TITAN IIICs were launched in the last half of 1969, but two more TITAN IIIC missions lifted off Complex 40 in 1970. The first of those two flights concluded the VELA detection satellite launch program by boosting ARPA's 11th and 12th VELA satellites into orbit on 8 April 1970. The other flight took place on November 6th, and it involved a classified DOD payload. Complex 40 continued to be used for classified missions during the 1970s and 1980s, but Complex 41 supported only a handful of TITAN III missions before it was deactivated at the end of 1977. Complex 41 was refurbished for the TITAN IV program during the last half of the 1980s, but its first TITAN IV launch did not take place until 14 June 1989 -- almost 12 years after it was used to launch the VOYAGER missions to the outer planets.[29]

The 6555th, Chapter IV, Section 5

Taking the High Ground: The 6555th's Role in Space through 1970

Organizational Changes 1965-1970

In the midst of so many different missions, the TITAN III Division's leadership and organization remained relatively stable. Lieutenant Colonel Ducote was promoted to Colonel in November 1965, and he continued to serve as the TITAN III Division Chief until his retirement on 30 November 1967. He was succeeded by Lieutenant Colonel Mack E. Baker, who had replaced Major Henry as Payloads Branch Chief in July 1965. Lieutenant Colonel Elmer T. Helms replaced Lieutenant Colonel Baker as Division Chief in 1969. Lieutenant Colonel John M. Kminek served as Acting Division Chief during the first half of 1970, before Lieutenant Colonel Julius R. Conti became the Division Chief later in the year. The only significant change in the Division's organization during this period came in May 1968, when the Division's activities were realigned to match functions in the Space and Missile Systems Organization (SAMSO). As a result of that realignment, the Division's Operations Branch and Test Support Branch were combined to form the Program Management Branch. The Systems Branch was renamed the Launch Vehicle Branch, and the Payloads Branch remained unchanged.[30]

On the other hand, change was evident in the Division's manpower tables long before the Manned Orbiting Laboratory program was cancelled in 1969. Though the TITAN III Division's strength rose to 36 officers, 76 airmen, and 19 civilians in 1965, it dropped to approximately 100 personnel in 1966. It fell to 90 officers, airmen and civilians in 1968, and, toward the end of 1970, the Division's manning authorizations were cut to 74. (Its actual manning was reduced below than number in 1971.) Manpower losses were even more severe in other parts of the 6555th, as more and more Air Force space and missile programs shifted to Vandenberg for operations in the mid-1960s.[31]

At its peak in January 1964, the 6555th Aerospace Test Wing had 144 officers, 573 airmen and 76 civilians assigned to its various activities. Unfortunately, the Wing also lost any chance for an ATLAS/AGENA blue suit launch capability in January 1964, and the TITAN II Weapons Division was discontinued at the end of June. The Wing's strength dropped to 147 officers, 398 airmen and 64 civilians by June 1964, and it fell to 104 officers, 324 airmen and 59 civilians by the end of June 1965. Further reductions were punctuated with the completion of the GEMINI Launch Vehicle Division's mission in November 1966, and the Wing's manning stood at 70 officers, 204 airmen and 47 civilians by the end of 1966. By the middle of 1967, the Wing's manning was down to 256 personnel, and 214 of those people were assigned to just two organizations -- the MINUTEMAN Division and the SLV-V (TITAN) Division. The Wing's complement of officers and airmen increased slightly in 1968, but overall manning slipped back to 262 officers, airmen and civilians by the end of 1969. Strength continued to decline as MINUTEMAN launch operations were wrapped up in 1970, and Lieutenant Colonel Conti's TITAN Systems Division had 78 of the Wing's remaining 154 personnel by the end of 1970.[32]

On 1 April 1970, the Wing was redesignated the 6555th Aerospace Test Group, and it was reassigned to the 6595th Aerospace Test Wing as one of the 6595th's three Groups. Though this realignment constituted a two-fold drop in status for the 6555th, it came as no surprise to officials who had been following the shift in Air Force operations toward Vandenberg in the 1960s. While Vandenberg and Cape Canaveral both supported Air Force ICBM test flights in the 1960s, the MINUTEMAN III R&D program was the last Air Force ballistic missile effort on the Eastern Test Range, and it would end in December 1970. Cape Canaveral remained an excellent launch base for manned space missions and deep space probes, but with the exception of the TITAN III program, the Cape's space launch facilities and programs were dominated by NASA. In contrast to the Eastern Test Range, the Western Test Range had long-term commitments to SAC and the Air Force for operational tests of the TITAN II, MINUTEMAN I, II and III, and later ballistic missile programs, and the Western Test Range was ideal for DOD launches that placed satellites into polar orbit. Most of those facts were known as early as 1967, when NASA space programs required 50 percent of the Eastern Test Range's total effort, and the U.S. Navy commanded another 30 percent of the Range's time as its principal customer for ballistic missile test support. When other users were considered, SAMSO actually provided only 11 percent of the Eastern Test Range's activity in 1967. On the other hand, SAMSO accounted for 45 percent of the workload at Vandenberg, and SAC required another 30 percent. NASA claimed the remaining 25 percent of Western Test Range's activity, mainly in the form of instrumentation ship support for the Apollo manned space program. By 1970, the Western Test Range's operations had become much more important to the Air Force than the Eastern Test Range's launch activities, so it was logical to give the 6555th a less prominent role.[33]

The 6555th Aerospace Test Group continued to serve with some distinction after 1970, but if one merely considers the 6555th's accomplishments through 1970, the Group clearly had many reasons to feel proud. For its outstanding missile safety record in 1961, the 6555th was awarded the USAF Missile Safety Plaque for the first time in history. It received that award two more times in 1966 and 1970. On 25 October 1962, the 6555th was awarded the first in a series of six Air Force Outstanding Unit Awards covering the Wing's efforts from 21 December 1959 to 31 March 1971. In October 1962, the Wing also received its first Group Achievement Award from NASA for support of Project MERCURY. A second NASA Group Achievement Award followed in December 1965 for the 6555th's role in the GEMINI VII/VI rendezvous mission, and the Wing also won the Theodore von Karman trophy in 1967 for support of Project GEMINI. While those awards paid tribute to the excellence, dedication and stamina of the 6555th, the most important reward was probably nothing more than a feeling shared by the 6555th and many of its contractors: as space age pioneers, they helped build the missile and launch vehicle programs which paved most of America's road into space.[34]

The 6555th:

Chapter Four Footnotes

satellite effort
The satellite program was transferred from the Wright Air Development Center to the Western Development Division during the first part of 1956, and the satellite development plan was approved by Major General Schriever on 2 April 1956. Research and Development costs were expected to total $114.7 million, with $39.1 million required through fiscal 1957.

executive officer
Colonel J. J. Cody, Jr. served in that capacity as point-of- contract from late May through the end of July 1958. He was succeeded by Colonel James E. Miller, who continued as the ARPA point-of-contact through late August 1958.

THOR-ABLE
The THOR ballistic missile was used as the first stage of the THOR-ABLE, THOR-ABLE I, THOR-ABLE II and THOR-ABLE-STAR. The ABLE second stage was an Aerojet-General booster rated at 7,700 pounds of thrust. The ABLE I added the Allegheny Ballistic Laboratory's 2,540-pound-thrust solid rocket as a third stage to the ABLE second stage. The THOR-ABLE II consisted of a THOR first stage and a modified Aerojet-General 10-40 second stage. Aerojet General's ABLE-STAR upper stage was designed to boost a 1,000-pound payload into a 300-mile orbit.

Pad 17A
The TRANSIT 1A navigational satellite was also launched from Pad 17A by Air Force contractors for ARPA on 17 September 1959, but the payload failed to achieve orbit.

MIDAS Project Division
The Division was phased out on 30 June 1960 after the Lockheed Aircraft Corporation completed its second and final MIDAS R&D launch from Complex 14 on 24 May 1960. Subsequent MIDAS launches were carried out at Vandenberg Air Force Base.

TS 609A
Under Major Howard M. Sloan, the TS 609A Operations Division was charged with developing a blue suit capability for a family of small solid propellant rocket test vehicles. The rockets were composed of several stages, and they would be used for a variety of space experiments in basic and

applied research for ARDC and NASA. The TS-609A program was given the name "BLUE SCOUT" in the latter half of 1960, but the various TS-609A launch vehicle configurations required more distinctive names to differentiate them. The BLUE SCOUT I was composed of an Aerojet-General solid rocket, a Thiokol TX-33 solid rocket, and an Allegheny Ballistic Laboratory ABL-X254 solid rocket. The SCOUT and the BLUE SCOUT II both included those rocket stages, plus an Allegheny Ballistic Laboratory ABL-X248 rocket. The BLUE SCOUT JUNIOR consisted of the TX-33, ABL-X254, an Aerojet-General AJ 10-41 rocket motor, and the NOTS 100A solid rocket. By the summer of 1960, at least 17 officers and airmen had attended factory orientation courses on the various rocket stages, and a blue suit on-the-job training program was underway at Complex 18 and Building 1366. When the BLUE SCOUT Operations Division merged with the BLUE SCOUT Project Division on 17 April 1961, it became the Operations Section under the BLUE SCOUT Branch. Major Sloan was transferred to the 6555th's Ballistic Missiles Office, and Lieutenant Colonel Jesse G. Henry became the Chief of the BLUE SCOUT Branch. By that time, 71 airmen were working for the BLUE SCOUT Operations Section, and many of them participated in the preparation and launch of BLUE SCOUT vehicles.

reorganization
The reorganization reflected Air Force Systems Command's management of space and missile activities under two separate intermediate headquarters: the Space Systems Division and the Ballistic Systems Division. The 6555th was assigned to the Ballistic Systems Division in April 1961, but it served both intermediate headquarters. On 1 July 1963, the 6555th was reassigned from the Ballistic Systems Division to the Space Systems Division with no change in manning or station. Once again, it served both intermediate headquarters.

renamed
The Space Programs Office became the Office of the Deputy for Space Systems, and the Ballistic Missiles Office became the Office of the Deputy for Ballistic Systems.

Technical Support Office
Following the redesignation of the Directorate of Support as the Technical Support Division in April 1961, the Technical Support Division's Engineering Branch was abolished in July, and its Missile and Ground Safety Branch was abolished in September 1961. The Division became the Technical Support Office subsequently, but it continued to: 1) coordinate technical requirements and facility activities, 2) consolidate budget requirements, 3) manage technical supplies and 4) provide staff logistics support for the entire Wing.

Lieutenant Colonel Harold A. Myers
Lieutenant Colonel Myers had been the Space Projects Division's Assistant Chief of Test Operations under Lieutenant Colonel Thomas W. Morgan, but he moved up to replace Morgan as Division Chief during the last half of 1960. Lieutenant Colonel Morgan spent the better part of a year at Air War College at Maxwell Air Force Base, Alabama before he returned to replace Lieutenant Colonel Erwin A. Meyer as Chief of the Space Programs Office on 15 July 1961. In the meantime, Lieutenant Colonel Myers continued as Space Projects Branch Chief following the

6555th's reorganization in April 1961.

Space Projects Branch
Though some of the Space Projects Branch's ATLAS-related functions were transferred to the new ATLAS Space Branch (formerly, the ATLAS Boosters Branch) as a result of this reorganization, the BLUE SCOUT Branch remained unaffected.

Convair launched
Convair also launched RANGER I and RANGER II from Complex 12, but neither of those flights met their objectives.

memorandum of agreement
Lieutenant General Schriever signed the AGENA B agreement in February 1961, making February 14th the effective date. The Memorandum of Agreement on Participation of the 6555th Test Wing in the CENTAUR R&D Test Program was signed by Dr. Kurt H. Debus, Director of the Marshall Space Flight Center's Launch Operations Directorate, and Colonel Paul R. Wignall, Commander of the 6555th Test Wing (Development) on 18 April 1961.

ATLAS "D" and AGENA B
The ATLAS "D" space booster was essentially a modified ATLAS ICBM. Like the ATLAS "D" series missile, the ATLAS space booster was 75 feet long and 10 feet in diameter. It was a kerosene-fueled vehicle powered by two (first-stage) 154,000-pound-thrust Rocketdyne vernier booster engines and a 57,000-pound-thrust (half stage) sustainer engine. The AGENA B upper stage was 21.6 feet long and five feet in diameter. It used Unsymmetrical Dimethylhydrazine (UDMH) for fuel and Inhibited Red Fuming Nitric Acid (IRFNA) as an oxidizer. The ATLAS "D" was used to launch MERCURY capsules, and the ATLAS D/AGENA B combination was used to launch other spacecraft. The combined weight of the ATLAS/AGENA-B vehicle (minus payload) was approximately 292,500 pounds.

NASA missions
MARINER I had to be destroyed several minutes after lift-off on July 22nd, but MARINER II was launched successfully on August 27th, and it sent back data from the vicinity of Venus on 14 December 1962. RANGER III and RANGER V failed to meet their primary objectives in January and October 1962, but RANGER IV impacted on the moon's surface successfully in April 1962. All three MERCURY missions in 1962 were successful, though the second manned flight in May -- AURORA 7 -- ended 250 miles downrange from the intended target area. AURORA 7's Lieutenant Colonel Scott Carpenter was picked up about three hours later.

ATLAS Space Branch
Lieutenant Colonel Hull became the Acting Deputy for Space Systems on 1 July 1962, and he was reassigned as Chief of the SLV II/IV Division (formerly the THOR/TITAN Space Branch) on 15 September 1962. Major John R. Mullady moved up from the ATLAS Space Branch's Test

Operations Section to succeed Lieutenant Colonel Hull as Chief of the ATLAS Space Branch in July 1962, and Mullady continued as Division Chief following the Branch's redesignation as the SLV-III Division on 1 October 1962.

LeDewey E. Allen, Jr.
Major Allen succeeded Major Mullady as Division Chief during the first half of 1963, and he continued in that capacity until the middle of January 1967. His successor, Lieutenant Colonel Alexander C. Kuras, moved into the Chief's office on 27 February 1967. Lieutenant Colonel Kuras was succeeded by Lieutenant Colonel Earl B. Essing in August 1967.

limited military launch capability
Following six launches in 1961, operations on Complex 18 were scaled back drastically. Toward the end of 1961, Aeroneutronic began providing only limited assistance to the BLUE SCOUT Branch via a Letter Contract.

DYNA SOAR
The X-20 DYNA SOAR project was an Air Force experimental effort to develop a manned space glider which could be boosted into orbit, maneuvered, and piloted back to earth. Plans for the program called for two unmanned and eight manned TITAN IIIC space flights with manned glider landings at Edwards Air Force Base. In 1960, DYNA SOAR contracts were awarded to Boeing as prime contractor, and to Minneapolis-Honeywell and RCA as sub-contractors for the guidance system and communications and tracking system, respectively. Approximately $400 million was spent on the project, but, at Secretary McNamara's request, it was stopped by President Johnson in December 1963 before any space flights were flown. The Manned Orbiting Laboratory (MOL) succeeded the DYNA SOAR as a TITAN IIIC mission, but it was cancelled in June 1969.

TITAN III
Though the TITAN IIIC's missions as a space booster for DYNA SOAR and the Manned Orbiting Laboratory (MOL) were eventually cancelled, the Air Force continued to develop the TITAN III to meet other military and non-military space mission requirements. The Titan IIIA and Titan IIIC both utilized a modified TITAN II ICBM first stage as their first stage core. That core stage was rated at 430,000 pounds of thrust at sea level, and it provided the TITAN IIIA with all its power at lift-off. Two 10-foot-diameter solid rocket boosters were attached to the basic "A" configuration to make the TITAN IIIC, and those five-segment solid rockets developed 2,314,000 pounds of thrust -- all the power the 1,300,000-pound TITAN IIIC needed to lift itself off the pad. (The TITAN IIIC's first stage core fired at an altitude of 28 nautical miles, later in the flight.) Both vehicles employed a liquid-fueled second stage (rated at 100,000 pounds of thrust) and a small, pressure-fed transtage (rated at 16,000 pounds of thrust) to place their payloads into orbit.

NASA space missions
THOR/ABLE STAR space vehicles were used for the Navy payloads, and THOR/DELTA space vehicles were launched by Air Force contractors for the following NASA missions in 1962: "Big

Shot I," TIROS 4, 5 and 6, OSO 1, ARIEL 1, TELSTAR I, ECHO A-12, EXPLORER 14 and 15, and RELAY 1.

NASA THOR/DELTA missions
Air Force contractors provided space booster support for the following NASA missions from Complex 17 in 1963: TIROS 7 and 8, TELSTAR II, EXPLORER 17 and 18, and SYNCOM A-25 and A-26.

transferred them to NASA
At the time of the transfer, NASA agreed to return Complex 17 and other THOR facilities to the Range at the end of NASA's DELTA program. In accordance with that agreement, NASA completed the transfer of those facilities back to the Air Force in October 1988.

Division
Lieutenant Colonel Albert became the Chief of the GEMINI Launch Vehicle Division in late July 1963, and Major Ausfahl moved over to become the Chief of the Division's Flight Test Operations Branch on 30 July 1963. Over the next five months, the Division grew from six officers, one airman and two civilians to 17 officers, eight airmen and five civilians. Twenty officers, 19 airmen and four civilians were assigned to the GEMINI Launch Division by the end of 1964, and the Division's complement of airmen increased to 35 as personnel were received from the BLUE SCOUT and SLV-III Division in 1965.

hurricanes
Only the second stage of the vehicle was taken down and stored in a hangar on 26 August 1964 in preparation for Hurricane Cleo, but the entire launch vehicle was dismantled and removed from Pad 19 in early September before Hurricane Dora passed over the Cape on September 9th. GEMINI Launch Vehicle 2 (GLV-2) was erected for the final time on Pad 19 on 14 September 1964.

ATLAS/AGENA target vehicles
Two ATLAS/AGENA target vehicles were expended in support of GT-9, the seventh manned TITAN II/GEMINI mission. Following an ATLAS/AGENA target vehicle flight failure on 17 May 1966, GT-9 was delayed until another ATLAS/AGENA could be launched from Complex 14 on 1 June 1966. Astronauts Thomas P. Stafford and Eugene A. Cernan were boosted into orbit from Complex 19 two days later.

TITAN II space boosters
Under the direction of AFSC's Space Systems Division, the Martin Company modified the basic TITAN II ICBM design to create the "man-rated" TITAN II/GEMINI launch vehicle design. The new vehicle's tanks were welded, inspected and tested at Martin's Denver Division before they were sent to Martin's Baltimore Division for systems integration and further testing. The TITAN II's liquid propellant rocket engines were built by the Aerojet-General Corporation, and they were

also sent to Baltimore for systems integration. General Electric provided radio command guidance for the vehicles, and the Burroughs Corporation was responsible for ground guidance computers. The Aerospace Corporation provided systems engineering and technical direction. Approximately 1,500 other companies provided parts for the TITAN II/GEMINI vehicle. As Chief of the GEMINI Launch Vehicle Division, Lieutenant Colonel Albert served as the Air Force test controller and ensured the TITAN II/GEMINI vehicle was ready to meet its mission on launch day. Martin's Gemini-Titan II Launch Operations Division erected the vehicle on Pad 19, checked it out and launched it.

agreement
Under the Webb-McNamara Agreement of 17 January 1963, NASA and the Department of Defense agreed to consider the Merritt Island Launch Area (north and west of Cape Canaveral) a NASA installation. The agreement also stated that the TITAN III site would be excluded from NASA's administration and considered part of the Atlantic Missile Range (later known as the Eastern Test Range) so it could be administered by the Air Force.

Integrate-Transfer-Launch (ITL) system
As the name suggests, the ITL system was designed to assemble, checkout and integrate the TITAN IIIC's major components before it transferred the TITAN IIIC booster to the pad for payload mating and launch operations. The ITL system consisted on a Vertical Integration Building (VIB) for erection of the Titan III's core stages, a Solid Motor Assembly Building (SMAB) where the solid booster segments were stacked, a railroad track network, a warehouse and various support buildings and storage areas.

TITAN IIIA
Though the first TITAN IIIA launch went well, the third stage of the vehicle malfunctioned, and the 3,750-pound dummy payload failed to achieve orbit. The second TITAN IIIA flight from Complex 20 was more successful, and that mission placed the vehicle's final stage and a 3,750-pound dummy payload into orbit on 10 December 1964. The third TITAN IIIA placed its transtage and a Lincoln Experimental Satellite (LES-1) into orbit on 11 February 1965. The fourth (and last) TITAN IIIA boosted LES-2 on a highly successful orbital mission on 6 May 1965. The first TITAN IIIC carried a 21,000-pound dummy payload into orbit.

personnel roster
In addition to those resources, Lieutenant Colonel Andrew Wright managed seven other officers, two non-commissioned officers and three civilians assigned to the 6555th's Systems Civil Engineering Office and the Civil Engineering Branch of the TITAN III Task Force.

three branches
The Systems Branch was responsible for the flight readiness of the TITAN IIIA and TITAN IIIC boosters and ground equipment. The Operations Branch managed construction, inspections and tests of ITL facilities. The Test Support Branch monitored test support requirements for current and

future TITAN III missions, and it updated requirements documents accordingly.

Vandenberg Air Force Base
The TITAN IIIB R&D program at Vandenberg Air Force Base was about one year behind the Cape Canaveral TITAN IIIA program, but it ended with the first TITAN IIIB/AGENA D launch at Vandenberg on 29 July 1966. The TITAN IIID was the first Vandenberg TITAN to be configured with solid rocket motors like the ones used on the TITAN IIIC. The 6595th Space Test Group launched the first TITAN IIID space booster from Vandenberg on 15 June 1971.

Manned Orbiting Laboratory
The Akwa-Downey Construction Company began building Complex 40's MOL Environmental Shelter shortly after the second TITAN IIIC mission in October 1965. The Shelter was ready for beneficial occupancy by the middle of June 1966, and American Machine and Foundry completed installation of the Shelter's work platforms two weeks later. As we noted earlier, a modified GEMINI capsule was launched from Complex 40 as part of an experimental mission on 3 November 1966. That capsule had been launched and recovered during the GEMINI 2 mission, but it was modified by the McDonnell Aircraft Company to perform a heat shield test for the MOL program. The modified capsule was installed in the MOL Environmental Shelter on 3 October 1966, and it was launched with the rest of the TITAN IIIC's payload one month later. Though the MOL program was terminated in 1969, the Shelter was modified to accept new TITAN payload fairings, and the "new" Shelter began supporting TITAN IIIC launch operations when flights resumed at Complex 40 in April 1970.

Complex 41
Complex 41 was used for a VIKING simulator mission and a HELIOS solar mission in 1974, two VIKING missions to Mars in 1975, another HELIOS mission in 1976, and two VOYAGER missions to the outer planets in 1977.

Space and Missile Systems Organization
The Ballistic Systems Division (BSD) and the Space Systems Division (SSD) were replaced by SAMSO on 1 July 1967, and the 6555th was reassigned from SSD to SAMSO on the same date.

assigned
During its first year under the Air Force Ballistic Missile Division, the 6555th increased its strength from 71 officers, 159 airmen and 21 civilians to 91 officers, 274 airmen and 59 civilians. The Wing had 96 officers, 393 airmen and 64 civilians by the end of December 1961, and 117 officers, 402 airmen and 73 civilians were assigned to various Wing agencies one year later. The 6555th's build-up to its peak strength started in early 1963, and the Wing had 124 officers, 580 airmen and 70 civilians working for it by the end of June 1963. Peak strength was achieved six months later.

drop in status
In the early 1960s, the 6555th and the 6595th had equal status under AFSC: the 6555th reported to

the Ballistic Systems Division (BSD), and the 6595th reported to Space Systems Division (SSD). Both wings reported to SSD later on, and they remained on an equal footing when BSD and SSD were inactivated and replaced by SAMSO. The reorganization in April 1970 involved the redesignation of the Air Force Western Test Range as the Headquarters, Space and Missile Center (SAMTEC), and the insertion of SAMTEC into the chain-of-command between the 6595th Aerospace Test Wing and SAMSO. Thus, the 6595th dropped one level in status, and the 6555th dropped to the level of the 6595th's other Groups -- the 6595th Space Test Group and the 6595th Missile Test Group.

dominated by NASA

Complex 11 was deactivated in 1964, and complexes 12, 14, 15, 18, 19 and 20 were either deactivated or sold for salvage in 1967. In 1966, NASA was assigned the second of two launch pads on Complex 17, and it controlled complexes 34, 36 and 37 in addition to its SATURN V launch complexes on Merritt Island. Complex 13 was transferred back to the Air Force from NASA in March 1968 to support a handful of ATLAS/AGENA missions, but it was one of only three Cape complexes devoted solely to Air Force space launch operations. The other two were TITAN III complexes 40 and 41.

principal customer for ballistic missile test

The Navy's launched its first POLARIS ballistic test missile on the Eastern Test Range on 13 April 1957, and it completed its 387th POLARIS flight there on 25 November 1967. Though the frequency of POLARIS launches dropped dramatically after 1971, Navy POSEIDON ballistic missile tests began on the Eastern Test Range on 16 August 1968. Navy ballistic missile tests constituted more than half of all the major launches on Eastern Test Range between 1966 and 1972, and the Navy continued to provide the lion's share of ballistic missile tests in the east throughout the 1970s.

Outstanding Unit Awards

The 6555th's first Outstanding Unit Award was presented for the Wing's efforts between 21 December 1959 and the end of May 1962. The second award recognized the Wing's achievements between 1 October 1962 and the end of September 1964, and the third award distinguished the 6555th for its actions from October 1964 to 31 December 1965. The Wing received the award for the fourth time for its activities in 1966, and it received its fifth Air Force Outstanding Unit Award for its performance from 1 January 1968 to 1 January 1970. The award was conferred a sixth time for the 6555th's work from 1 April 1970 to 31 March 1971. The Group received the award for the seventh time in 1974, for the eighth time in 1978, for the ninth time in 1982 and for the tenth time in October 1990.

The 6555th

Chapter Four Endnotes

1. Perry, Origins, pp. 9, 12, 20, 34.

2. Ibid., pp. 35, 41, 44, 53, 55, 56, 57.

3. ARDC History, 1 July - 31 December 1959, "Foreword" and "Chronology"; AFSCF History Office, " AFSCF History Brief and Chronology, 1954 - 1981," p. 2; Ley, Men in Space, p. 362; 6555th ABG History Office, "Chronology of Atlantic Missile Range and Air Force Missile Test Center, 1938 - 1959," p. 115; General Directive Number 1, NASA, "Proclamation on Organization of the National Aeronautics and Space Administration," 25 September 1958; Executive Order Number 10783, 1 October 1958.

4. ARDC History, 1 July - 31 December 1959, p. II-39

5. Ibid., pp. I-9, I-27, I-52; ARDC History, 1 January - 30 June 1959, p. II-30; ARDC History, 1 January - 31 March 1961, p. I-33.

6. History of the Assistant Commander for Missile Tests, AFBMD, 1 June - 20 December 1959, ("Organization and Mission" and "Personnel"); 6555th Test Wing (Development) History, 21 December 1959 - 31 March 1960, DWTI Historical Section ("Introduction," "Physical Facilities" and "Missile Test Activities"); 6555th Test Wing (Development) History, 1 April - 30 June 1960, DWTI Historical Section ("Missile Test Activities"); AFMTC History, 1 January - 30 June 1959, pp. 169, 170; AFMTC History, 1 July - 31 December 1958, pp. 181, 182; AFMTC History, 1 July - 31 December 1959, pp. 179, 180.

7. 6555th Test Wing (Development) History, 21 December 1959 - 31 March 1960, DWOS Historical Section ("Introduction" and "Mission"), DWSF Historical Section ("New Construction"), DWTR Historical Section ("Introduction," Physical Facilities" and "Test Program"), and DWTS Historical Section ("Introduction," "Physical Facilities" and "Contract"); 6555th Test Wing (Development) History, 1 April - 30 June 1960, DWOS Historical Section ("Introduction" and "Missile Test Activities"), DWTI Historical Section ("Introduction") and DWTR Historical Section ("Organization and Functions" and "Test Activities"); 6555th Test Wing (Development) History, 1 January - 30 June 1961, DWSF Historical Section ("Construction") and DWZS Historical Section ("Organization and Functions" and "Personnel"); 6555th Test Wing (Development) History, 1 July

- 31 December 1960, DWTI Historical Section ("Physical Facilities" and "Missile Test Activities") and DWTS Historical Section ("Missile Test Activities").

8. 6555th Test Wing (Development) History, 1 January - 30 June 1961, DWS Historical Section ("Introduction"), DWT Historical Section ("Organization and Mission"), DWTC Historical Section ("Introduction"), DWZC Historical Section ("Organization and Mission"), DWZI Historical Section ("Introduction"), DWZS Historical Section ("Organization and Functions") and DWZT Historical Section ("Introduction"); 6555th Aerospace Test Wing History, 1 July - 31 December 1961, "Introduction," DWT Historical Section ("Organization and Mission") and DWZ Historical Section ("Introduction").

9. 6555th Test Wing (Development) History, 1 January - 30 June 1961, DWZI Historical Section ("Introduction," "Physical Facilities" and "Personnel"); 6555th Test Wing (Development) History, 1 July - 31 December 1960, DWTI Historical Section ("Personnel"); 6555th Test Wing (Development) History, 21 December 1959 - 31 March 1960, DWTI Historical Section ("Personnel"); 6555th Aerospace Test Wing History, 1 July - 31 December 1961, DWZ Historical Section ("Personnel") and DWZI Historical Section ("Introduction").

10. Whipple, "Index, July 1950 - June 1960," pp. 27-3, 27-5; Whipple "Index, July 1960 - June 1961," pp. 11-6, 11-8, A-1-1; Whipple, "Index, July 1961 - June 1962," pp. 2, 3, 5; 6555th Test Wing (Development) History, 1 January - 30 June 1961, DWZ Historical Section ("Problem Areas") and DWZS Historical Section ("Missile Test Activities"); 6555th Aerospace Test Wing History, 1 July -31 December 1961, DWZS Historical Section ("Missile Test Activities"); NASA and 6555th Aerospace Test Wing, "Memorandum of Agreement on Participation of the 6555th Test Wing (Development) in the Centaur R&D Flight Test Program," 18 April 1961; NASA, "National Aeronautics and Space Administration Agena B Launch Vehicle Program Management Organization and Procedures," 14 February 1961.

11. NASA, "National Aeronautics and Space Administration Agena B Launch Vehicle Program Management Organization and Procedures," 14 February 1961, pp. 1, 2, 3, 6, 7, 8.

12. NASA and 6555th Aerospace Test Wing, "Memorandum of Agreement on Participation of the 6555th Test Wing (Development) in the Centaur R&D Flight Test Program," 18 April 1961, pp. 1, 2, 3, and Addendum.

13. 6555th Test Wing (Development) History, 1 January - 30 June 1961, DWZC Historical Section ("Personnel"); 6555th Aerospace Test Wing History, 1 July - 31 December 1961, DWZC Historical Section ("Organization and Mission," "Missile Test Activities" and "Activities") and DWZT Historical Section ("Introduction"); Letter, Colonel Paul R. Wignall, 6555th ASTW/CC, to Dr. Kurt Debus, Director LOD, "Complex 36A Management Control," 19 October 1961.

14. 6555th Aerospace Test Wing History, 1 January - 30 June 1962, DWZC Historical Section

("Activities"); 6555th Aerospace Test Wing History, 1 July - 31 December 1962, DWZ Historical Section ("Key Personnel and Positions"), DWZC Historical Section ("Organization and Mission," "Personnel" and "Activities"), and DWZT Historical Section ("Introduction"); 6555th Aerospace Test Wing History, 1 January - 30 June 1963, DWTC Historical Section ("Personnel Losses") and DWZC Historical Section ("Organization and Mission, "Strength Resume" and "Activities"); 6555th Aerospace Test Wing History, 1 January - 30 June 1964, DWZC Historical Section ("Program Activities").

15. 6555th Aerospace Test Wing History, 1 January - 30 June 1967, DWC Historical Section ("Personnel"); 6555th Aerospace Test Wing History, 1 July - 31 December 1963, DWZC Historical Section ("Activities"); 6555th Aerospace Test Wing History, 1 January - 30 June 1964, DWZC Historical Section ("Activities"); 6555th Aerospace Test Wing History, 1 July - 31 December 1964, DWZ Historical Section ("Activities"); 6555th Aerospace Test Wing History, 1 January - 30 June 1965, DWC Historical Section ("Division Test Activities"); 6555th Aerospace Test Wing History, 1 July - 31 December 1965, DWC Historical Section ("Guidance Activities"); 6555th Aerospace Test Wing History, 1 January - 30 June 1966, DWC Historical Section ("Division Test Activities"); 6555th Aerospace Test Wing History, 1 July - 31 December 1966, DWC Historical Section ("Division Test Activities"); Crespino, "Launches," p. 4; Whipple, "Index, July 1963 - June 1964," p. 2; 6555th Aerospace Test Wing History, 1 January - 30 June 1967, DWC Historical Section ("Mission, Objectives and Organization"); 6555th Aerospace Test Wing History, 1 July - 31 December 1967, DWC Historical Section ("Mission, Objectives and Organization"); 6555th Aerospace Test Wing History, 1 January - 30 June 1968, DWC Historical Section ("Division Test Activities"); Crespino, "Launches," p. 4; 6555th Aerospace Test Wing History, 1 July - 31 December 1969, DWC Historical Section ("Strength Resume"); 6555th Aerospace Test Group History, 1 January - 30 June 1970, ATLAS Systems Division Historical Section ("ATLAS Systems Division Test Activities").

16. 6555th Aerospace Test Wing History, 1 July - 30 June 1961, DWZS Historical Section ("Missile Test Activities"); 6555th Aerospace Test Wing History, 1 January - 30 June 1962, DWZS Historical Section ("Personnel" and "Missile Test Activities"); 6555th Aerospace Test Wing History, 1 July-31 December 1962, DWZS Historical Section ("Organization" and "Key Personnel Changes"); 6555th Aerospace Test Wing History, 1 January - 30 June 1963, DWZS Historical Section ("Facilities," "Duty Assignments" and "Training"); Crespino "Launches," pp. 6, 30; 6555th Aerospace Test Wing History, 1 July - 31 December 1963, DWZS Historical Section ("Test Program"); 6555th Aerospace Test Wing History, 1 January - 30 June 1964, DWZS Historical Section ("Test Program," "Duty Assignments" and "Problem Areas"); 6555th Aerospace Test Wing History, 1 January - 30 June 1965, DWS Historical Section ("Missile Test Activities" and "Personnel Assignments"); 6555th Aerospace Test Wing History, 1 July - 31 December 1965, "Organization."

17. Headquarters Space Division History Office, "Space and Missile Systems Organization: A Chronology, 1954 - 1979," p. 198; 6555th Aerospace Test Wing History, 1 July - 31 December 1961, DWZT Historical Section ("Introduction"); 6555th Aerospace Test Wing History, 1 January

- 30 June 1962, DWZT Historical Section ("Introduction," "Physical Facilities" and "Missile Test Activities"); 6555th Aerospace Test Wing History, 1 July - 31 December 1962, DWZB Historical Section, ("Introduction," "X-20 Activities" and "Program 624A Activities") and DWZT Historical Section ("Organizational Structure" and "Missile Test Activities"); Handbook, Martin Company, "USAF TITAN III Standard Space Launch System," 3rd Edition, undated, pp. A-2, A-3; 6555 Aerospace Test Wing History, 1 January - 30 June 1963, DWZB Historical Section ("Introduction," "Strength Resume" and "Physical Facilities"), DWZG Historical Section ("Introduction" and "Key Personnel") and DWZT Historical Section ("Introduction," "Key Personnel" and "Missile Test Activities"); 6555th Aerospace Test Wing History, 1 July - 31 December 1963, DWZT Historical Section ("Strength Resume," "Key Personnel and Positions" and "Missile Test Activities"); Whipple, "Index, July 1961 - June 1962," p. 10; Whipple, "Index, July 1962 - June 1963," pp. 5, 6; Article, "MOL is Replacing Dyna-Soar to Test Man's Ability in Space," St. Louis Post Dispatch, 11 December 1963; Article, "MOL Replaces Dead Dyna-Soar," Orlando Sentinel, 11 December 1963.

18. 6555th Aerospace Test Wing History, 1 July - 31 December 1963, DWZT Historical Section, ("Strength Resume" and "Key Personnel and Positions"); 6555 Aerospace Test Wing History, 1 January - 30 June 1964, DWZT Historical Section ("Missile Test Activities"); 6555th Aerospace Test Wing History, 1 July - 31 December 1964, DWZ Historical Section ("Program Activities"); 6555th Aerospace Test Wing History, 1 January - 30 June 1965, DWT Historical Section ("Activities"); Whipple, "Index, July 1963 - June 1964," p. 29; Whipple, "Index, July 1964 - June 1965," p. 31.

19. 6555th Aerospace Test Wing History, 1 January - 30 June 1965, DWT Historical Section ("Introduction"); USAF and NASA, "Agreement Between USAF and NASA for Transition of the Delta Space Launch Vehicle Program," signed 1 July 1988; Letter, Mr. James D. Phillips to ESMC/DER, "Transfer of Real Property (Delta Program)," 18 August 1988; 6550th ABG/DER, "Weekly Activity Report," 4 October 1988.

20. 6555th Aerospace Test Wing History, 1 July - 31 December 1963, DWZG Historical Section ("Test Activities"); Whipple, "Index, July 1963 - June 1964," p. 43; 6555th Aerospace Test Wing History, 1 July - 31 December 1964, DWD Historical Section ("Test Activities"); 6555th Aerospace Test Wing History, 1 January - 30 June 1965, DWD Historical Section ("Personnel" and "Test Activities"); Handbook, The Martin Company, " GEMINI - TITAN II Air Force Launch Vehicle Press Handbook," 2 February 1967, pp. 1-1, 1-2, 7- 2.

21. Handbook, The Martin Company, "GEMINI - TITAN II Air Force Launch Vehicle Press Handbook," 2 February 1967, pp. 1-1, 1-2, 1-3, A-1, A-2, and D-1 through D-19; 6555th Aerospace Test Wing History, 1 January - 30 June 1965, DWD Historical Section ("Test Activities"), 6555th Aerospace Test Wing History, 1 July - 31 December 1965, DWC Historical Section ("SLV III Division Test Activities") and DWD Historical Section ("Test Activities"); 6555th Aerospace Test Wing History, 1 January - 30 June 1966, DWC Historical Section ("SLV III Division Test Activities") and DWD Historical Section ("Test Activities"); 6555th Aerospace Test

Wing History, 1 July - 31 December 1966, "Organization," DWC Historical Section ("SLV III Division Test Activities") and DWD Historical Section ("Strength Resume" and "Test Activities"); 6555th Aerospace Test Wing History, 1 January - 30 June 1967, "Strength Resume" and "Missile and Space Program Activities."

22. Robert F. Piper, "History of TITAN III," June 1964, pp. 17, 19, 30, 37, 39, 40, 41.

23. AFETR History, 1964, Volume II, p. 65; DOD and NASA, "Agreement between The Department of Defense and The National Aeronautics & Space Administration Regarding Management of The Atlantic Missile Range of DOD and The Merritt Island Launch Area of NASA," signed 17 January 1963, paragraphs IB2 and IIIB; Handbook, The Martin Company, "USAF TITAN III Standard Space Launch System," 3rd Edition, undated, p. C-4.

24. Headquarters Space Division History Office, "Space and Missile Systems Organization: A Chronology, 1954 - 1979," pp. 198, 243, 273, 274; Handbook, The Martin Company, "USAF TITAN III Standard Space Launch System," 3rd Edition, undated, p. C-4; 6555th Aerospace Test Wing History, 1 January - 30 June 1963, DWF Historical Section ("Facilities Activity") and DWZB Historical Section ("Program 624A Activities"); Aerospace Corporation, "Titan III ITL Systems Launch Facilities at Cape Kennedy," 31 March 1965, p. 1; 6555th Aerospace Test Wing History, 1 July - 31 December 1963, DWF Historical Section ("Facilities Activity"); 6555th ASTW/ DWF, "Commander's Fact Book," 1 March 1965, pp. 1, 2; 6555th Aerospace Test Wing History, 1 January - 30 June 1965, DWF Historical Section ("Facility Activities").

25. 6555th Aerospace Test Wing History, 1 January - 30 June 1963, DWZB Historical Section ("Organization and Mission," "Strength Resume" and "Program 624 Activities"); AFETR History, 1964, Volume II, pp. 70, 71, 72; 6555th Aerospace Test Wing History, 1 January - 30 June 1964, DWZB Historical Section ("Physical Facilities"); Whipple, "Index, July 1964 - June 1965," pp. 27, 29; 6555th Aerospace Test Wing History, 1 January - 30 June 1965, DWB Historical Section ("Organization and Mission" and "Physical Facilities"); 6555th Aerospace Test Wing History, 1 July - 31 December 1965, DWB Historical Section ("Organization and Mission" and "Division Activities").

26. 6555th Aerospace Test Wing History, 1 July - 31 December 1963, DWZB Historical Section ("Introduction," "Organizational Structure" and "Strength Resume"); 6555th Aerospace Test Wing History, 1 January - 30 June 1964, DWZB Historical Section ("Introduction" and "Personnel"); 6555 Aerospace Test Wing History, 1 July - 31 December 1964, DWB Historical Section ("Organization and Mission" and "Personnel"); Headquarters Space Division History Office, "Space and Missile Systems Organization: A Chronology, 1954 - 1979," p. 174, 212; 6555th Aerospace Test Wing History, 1 January - 30 June 1965, DWB Historical Section ("Division Activities").

27. 6555th Aerospace Test Wing History, 1 July - 31 December 1965, DWB Historical Section

("Introduction" and "Physical Facilities"); 6555th Aerospace Test Wing History, 1 January - 30 June 1966, DWB Historical Section ("Introduction"); 6555th Aerospace Test Wing History, 1 July - 31 December 1966, DWB Historical Section ("Introduction"); Whipple, "Index, July 1965 - June 1966," p. 23; Whipple, "Index, July 1966 - June 1967," p. 21.

28. Whipple, "Index, July 1966 - June 1967," pp. 21, 22; Whipple, "Index, July 1967 - June 1968," p. 20; 6555th Aerospace Test Wing History, 1 July - 31 December 1965, DWB Historical Section ("Physical Facilities"); 6555th Aerospace Test Wing History, 1 January - 30 June 1966, DWB Historical Section ("Physical Facilities"); 6555th Aerospace Test Wing History, 1 July - 31 December 1966, DWB Historical Section ("Physical Facilities"); 6555th Aerospace Test Wing History, 1 January - 30 June 1967, DWB Historical Section ("Physical Facilities" and "Missile Test Activities"); 6555th Aerospace Test Wing History, 1 July - 31 December 1967, DWB Historical Section ("Division Activities"); 6555th Aerospace Test Wing History, 1 January - 30 June 1968, DWB Historical Section ("Division Activities"); 6555th Aerospace Test Wing History, 1 July - 31 December 1968, DWB Historical Section ("Division Activities"); 6555th Aerospace Test Wing History, 1 July - 31 December 1969, DWB Historical Section ("Division Activities"); 6555th Aerospace Test Wing History, 1 June - 31 December 1970, DWB Historical Section ("Facility Activities"); Whipple, "Index, July 1968 - June 1969," p. 22.

29. Whipple "Index, July 1969 - June 1970," p. 27; 6555th Aerospace Test Wing History, 1 July - 31 December 1969, DWB Historical Section ("Division Activities"); 6555th Aerospace Test Wing History, 1 January - 30 June 1970, TS Historical Section ("Division Activities");6555th Aerospace Test Wing History, 1 July - 31 December 1970, TS Historical Section ("Division Activities"); Crespino, "Launches," p. 56.

30. 6555th Aerospace Test Wing History, 1 July - 31 December 1965, DWB Historical Section ("Personnel"); 6555th Aerospace Test Wing History, 1 July - 31 December 1967, "Introduction" and DWB Historical Section ("Personnel"); 6555th Aerospace Test Wing History, 1 January - 30 June 1968, DWB Historical Section ("Mission" and "Organizational Structure"); 6555th Aerospace Test Wing History, 1 July - 31 December 1969, DWB Historical Section ("Personnel"); 6555th Aerospace Test Wing History, 1 January - 30 June 1970, TS Historical Section ("Personnel"); 6555th Aerospace Test Wing History, 1 July - 31 December 1970, TS Historical Section ("Personnel").

31. 6555th Aerospace Test Wing History, 1 July - 31 December 1965, DWB Historical Section ("Promotions"); 6555th Aerospace Test Wing History, 1 January - 30 June 1966, DWB Historical Section ("Strength Resume"); 6555th Aerospace Test Wing History, 1 July - 31 December 1966, DWB Historical Section ("Strength Resume"); 6555th Aerospace Test Wing History, 1 January - 30 June 1967, DWB Historical Section ("Strength Resume"); 6555th Aerospace Test Wing History, 1 July - 31 December 1967, DWB Historical Section ("Strength Resume"); 6555th Aerospace Test Wing History, 1 January - 30 June 1968, DWB Historical Section ("Strength Resume"); 6555 Aerospace Test Wing History, 1 July - 31 December 1968, DWB Historical Section ("Strength Resume"); 6555th Aerospace Test Wing History, 1 July - 31 December 1969,

DWB Historical Section ("Strength Resume"); 6555th Aerospace Test Wing History, 1 January - 30 June 1970, TS Historical Section ("Strength Resume"); 6555th Aerospace Test Wing History, 1 July - 31 December 1970, TS Historical Section ("Strength Resume"); 6555th Aerospace Test Wing History, 1 January - 30 June 1971, TS Historical Section ("Strength Resume");6555th Aerospace Test Wing History, 1 July - 31 December 1971, TS Historical Section ("Strength Resume").

32. 6555th Test Wing (Development) History, 21 December 1959 - 31 March 1960, DWA Historical Section ("Assigned Personnel Strength Summary"); 6555th Test Wing (Development) History, 1 July - 31 December 1960, DWP Historical Section ("Wing Strength Resume"); 6555th Test Wing (Development) History, 1 July - 31 December 1961, DWP Historical Section ("Wing Strength Resume"); 6555th Aerospace Test Wing History, 1 July - 31 December 1962, Chapter II ("Wing Strength Resume"); 6555th Aerospace Test Wing History, 1 January - 30 June 1963, Chapter II ("Wing Strength Resume"); 6555th Aerospace Test Wing History, 1 July - 31 December 1963, Chapter II ("Wing Strength Resume"); 6555th Aerospace Test Wing History, 1 January - 30 June 1964, Chapter I ("Organization") and Chapter II ("Wing Strength Resume"); 6555th Aerospace Test Wing History, 1 January - 30 June 1965, Chapter II ("Wing Strength Resume"); 6555th Aerospace Test Wing History, 1 July - 31 December 1966, Chapter II ("Wing Strength Resume"); 6555th Aerospace Test Wing History, 1 January - 30 June 1967, Chapter II ("Wing Strength Resume"), DWB Historical Section ("Strength Resume") and DWQ Historical Section ("Strength Resume"); 6555th Aerospace Test Wing History, 1 January - 30 June 1968, Chapter II ("Wing Strength Resume"); 6555th Aerospace Test Wing History, 1 July - 31 December 1968, Chapter II ("Wing Strength Resume"); 6555th Aerospace Test Wing History, 1 July - 31 December 1969, Chapter II ("Wing Strength Resume"); 6555th Aerospace Test Wing History, 1 January - 30 June 1970, Chapter II ("Wing Strength Resume"), MT Historical Section ("Strength Resume") and TS Historical Section ("Strength Resume"); 6555th Aerospace Test Wing History, 1 July - 31 December 1970, Chapter II ("Wing Strength Resume"), MT Historical Section ("Strength Resume") and TS Historical Section ("Strength Resume").

33. Special Order G-34, HQ AFSC, 30 March 1970; Message, HQ AFSC to HQ SAMSO, "Establishment of SAMTEC," 120019Z March 1970; Article, "6555th Test Wing Renamed in SAMSO Reorganization," The Missileer, 10 April 1970; Article, "Efficiency Dictates Test Center's Formation," Astronews, 17 April 1970; Headquarters Space Division History Office, "Space and Missile Systems Organization: A Chronology, 1954 - 1979," p. 12; Headquarters AFSC History Office, "Organizational Charts of the Air Force Systems Command, 1950 To Present," November 1987, pp. II-10 through II-14, II-16, II-19; 6555th Aerospace Test Wing History, 1 January - 30 June 1970, Chapter I ("Organization and Mission"); Whipple, "Index, July 1950 - June 1960," p. 20-2; Crespino, "Launches," introductory summary and pp. 27, 44; Study, HQ SAMTO, "Centroid Study Panel on Test Wings and Ranges," 21 November 1967; AFETR History, 1965 - 1966, Volume II, pp. 7, 9, 10, 11; AFETR History, 1 January 1967 - 30 June 1968, Volume I, Part 2, pp. 319 through 322; AFETR History, Fiscal 1969, Volume I, Part 2, p. 271; AFETR History, FY 1970, Volume I, Part 2, p. 300; AFETR History, Fiscal Year 1971, Volume I, Part 2, p. 288.

34. Pamphlet, 6555th Aerospace Test Wing, "The Story of the 6555th Aerospace Test Wing," o/a July 1966, pp. 21, 23, 26, 30, 33; 6555th Aerospace Test Group History, 1 July - 31 December 1970, Chapter III ("Honors and Awards"); 6555th Aerospace Test Group History, 1 January - 30 June 1974, Chapter III ("Honors and Awards"); 6555th Aerospace Test Group History, 1 January - 30 June 1976, Frontispiece; 6555th Aerospace Test Group History, January - September 1978, Chapter II ("Awards"); AFSC History, 1 July - 31 December 1961, Volume I, p. 91; 6555th Aerospace Test Wing History, 1 July - 31 December 1967, Chapter III ("Missile and Space Program Activities"); Article, "Test Group Named Outstanding Unit," The Missileer, 20 November 1970.

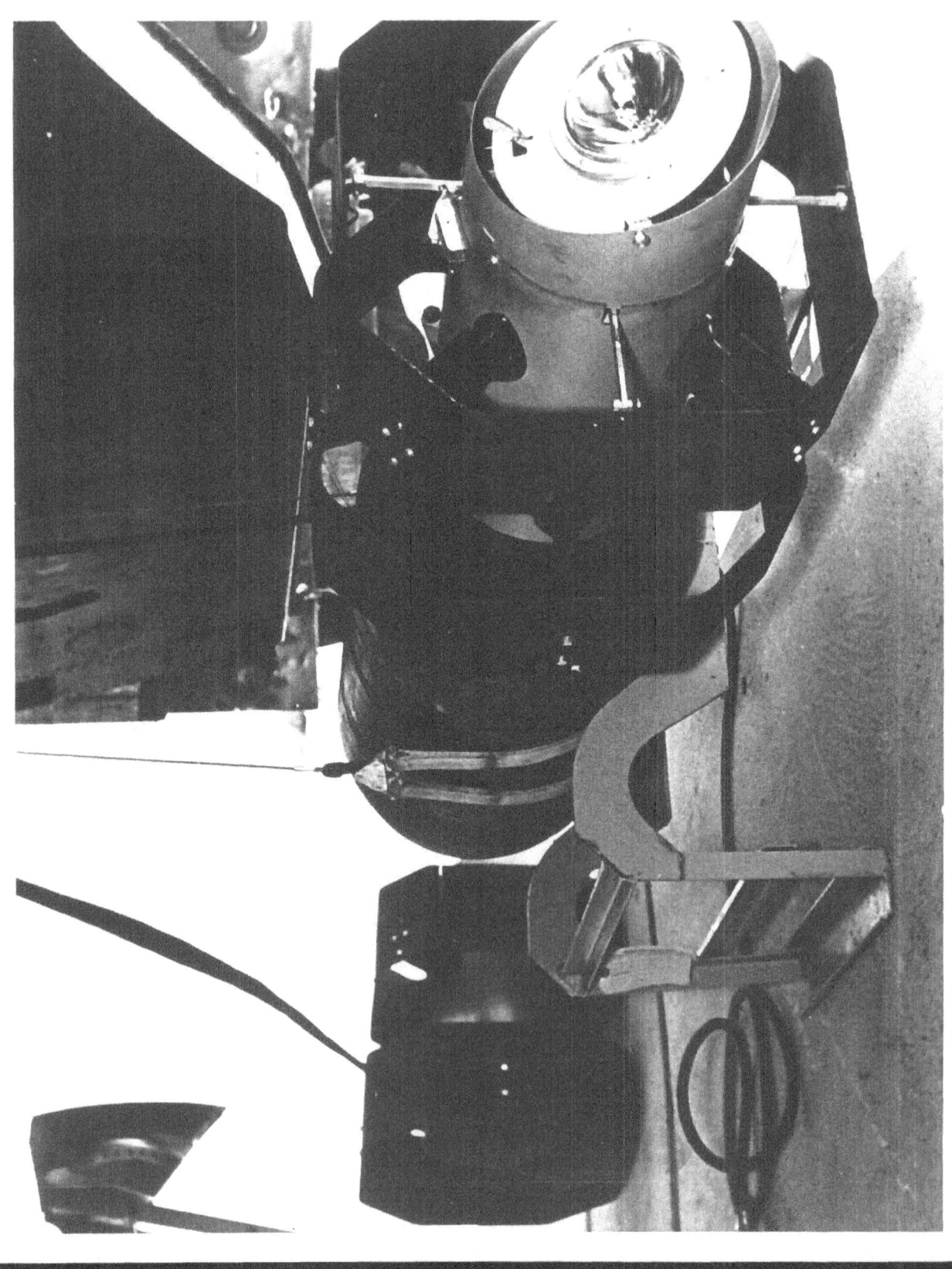

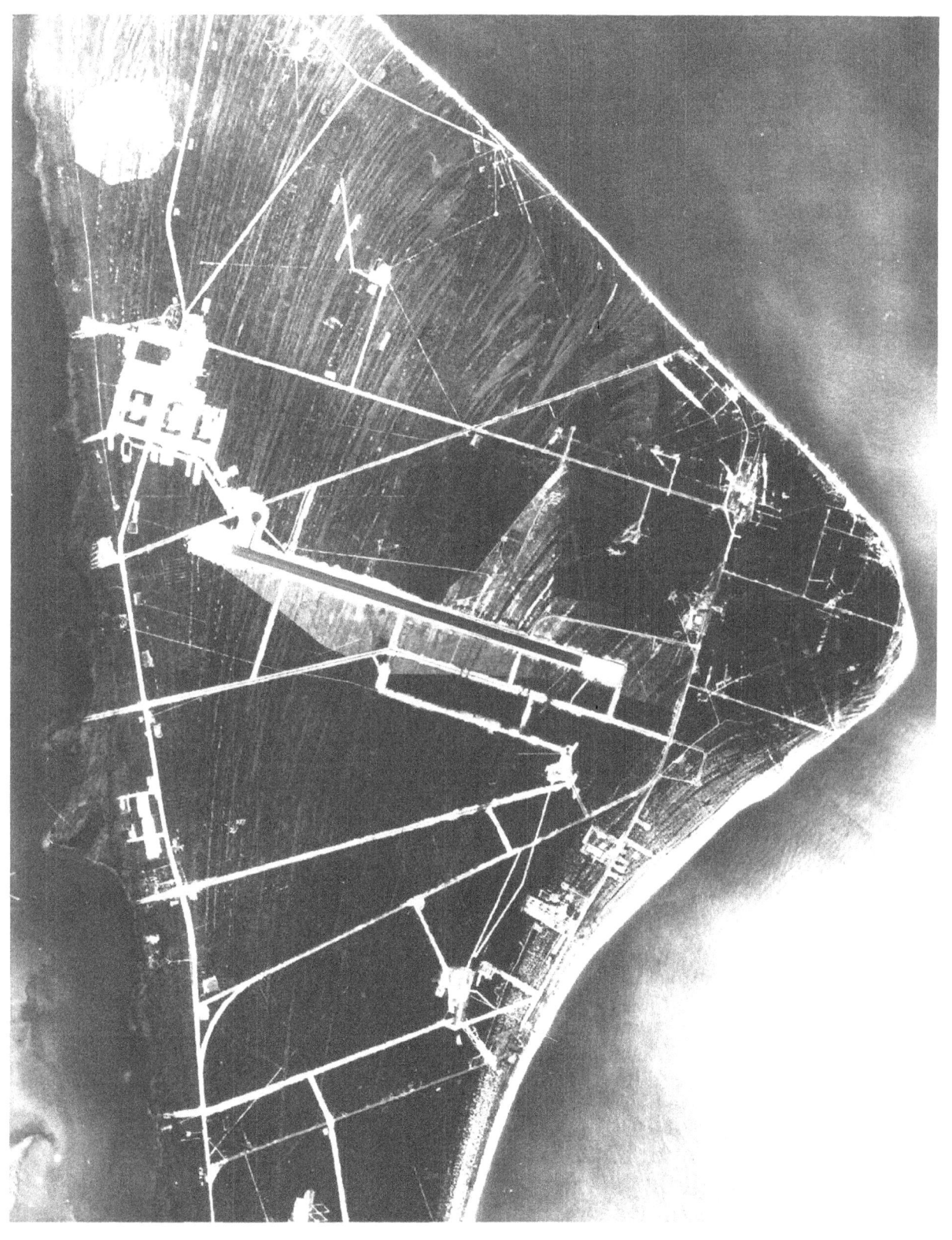

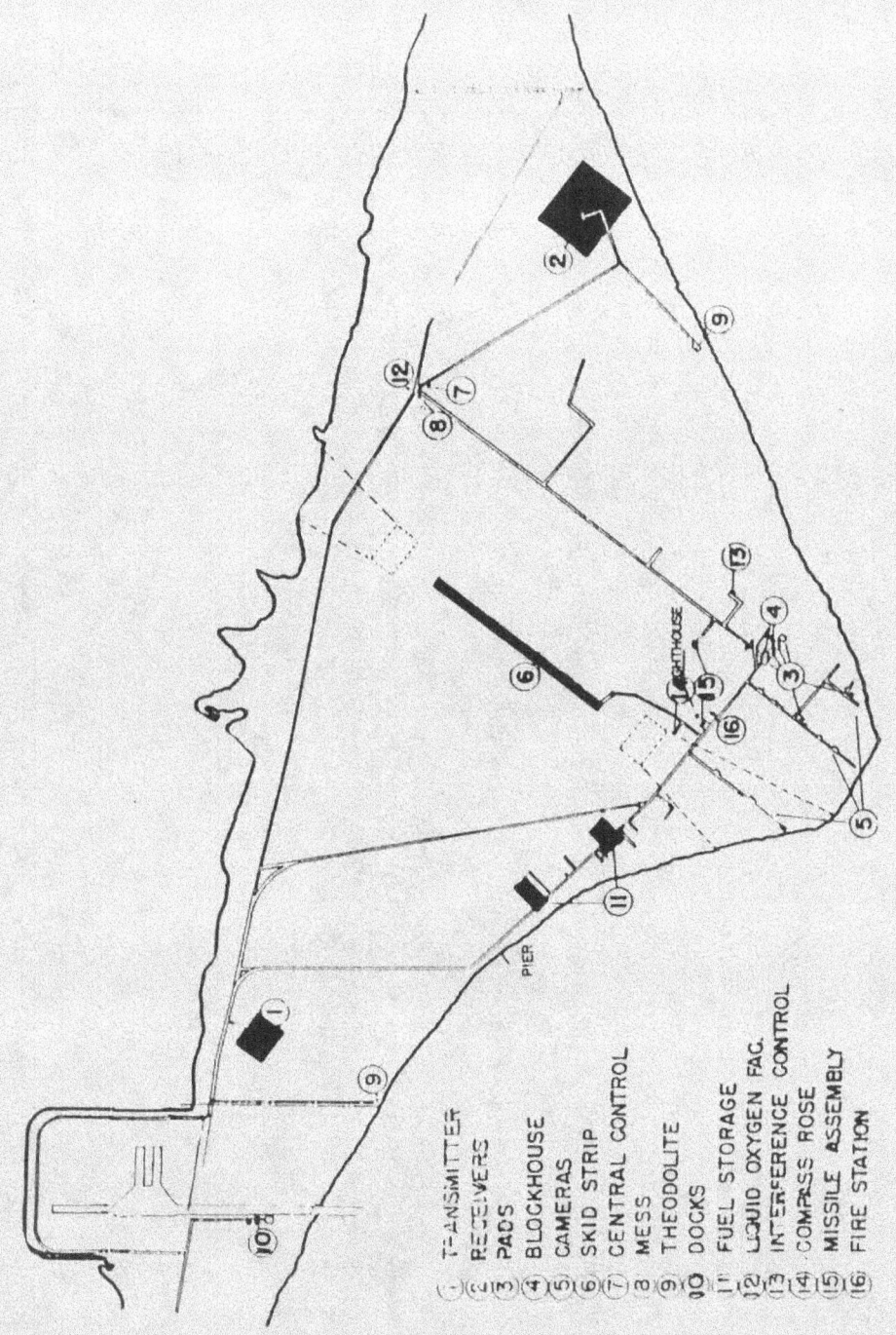

TRANSPORT VEHICLE

FIGURE 1 FRONT VIEW

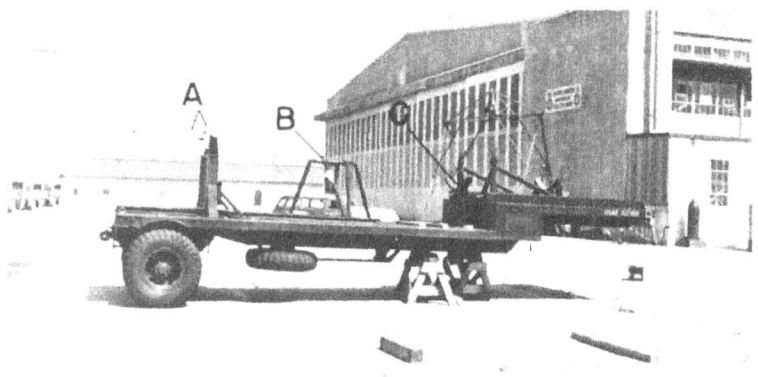

FIGURE 2 SIDE VIEW

A. MISSILE CENTER SECTION SUPPORT ARMS.
B. WING SUPPORT BRACKET.
C. MISSILE CENTER SECTION REAR CRADLE.
D. TAIL SECTION SUPPORT ARMS.
E. TAIL SECTION REAR CRADLE.

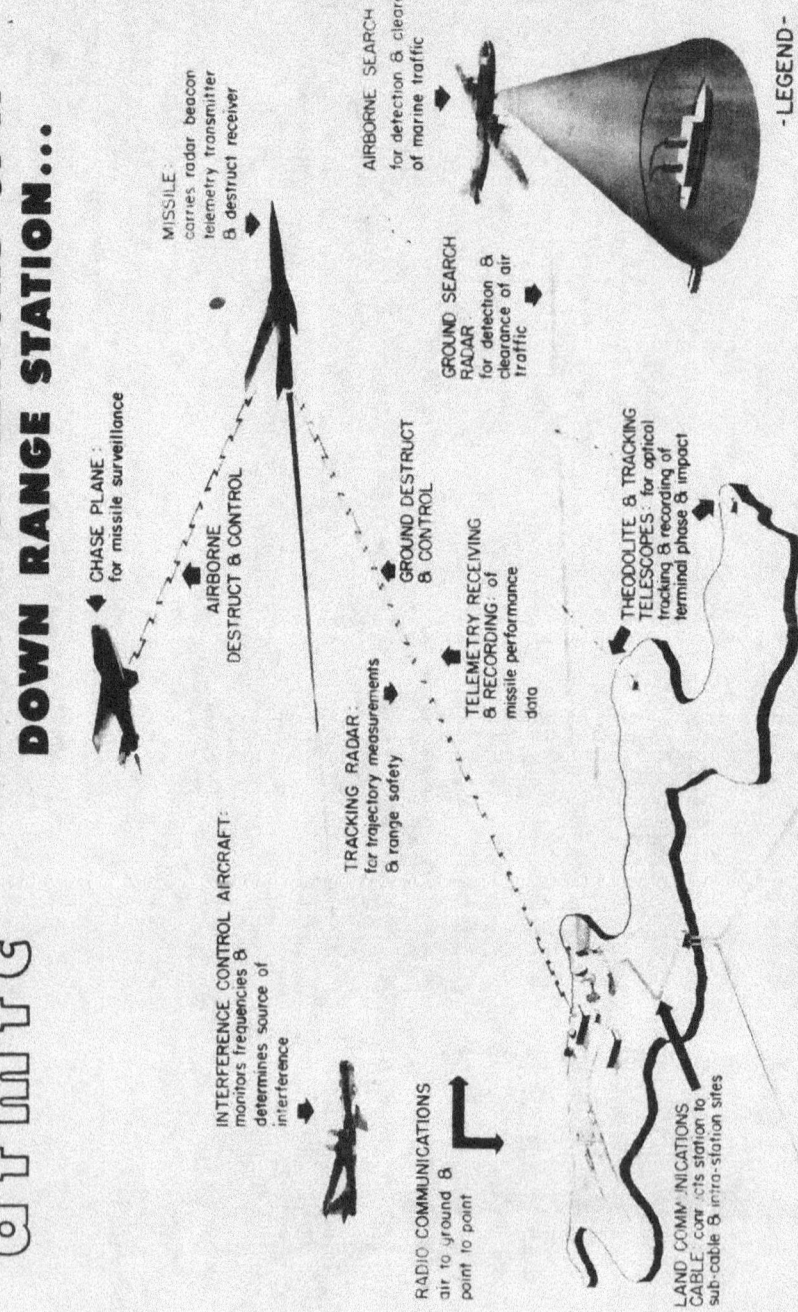

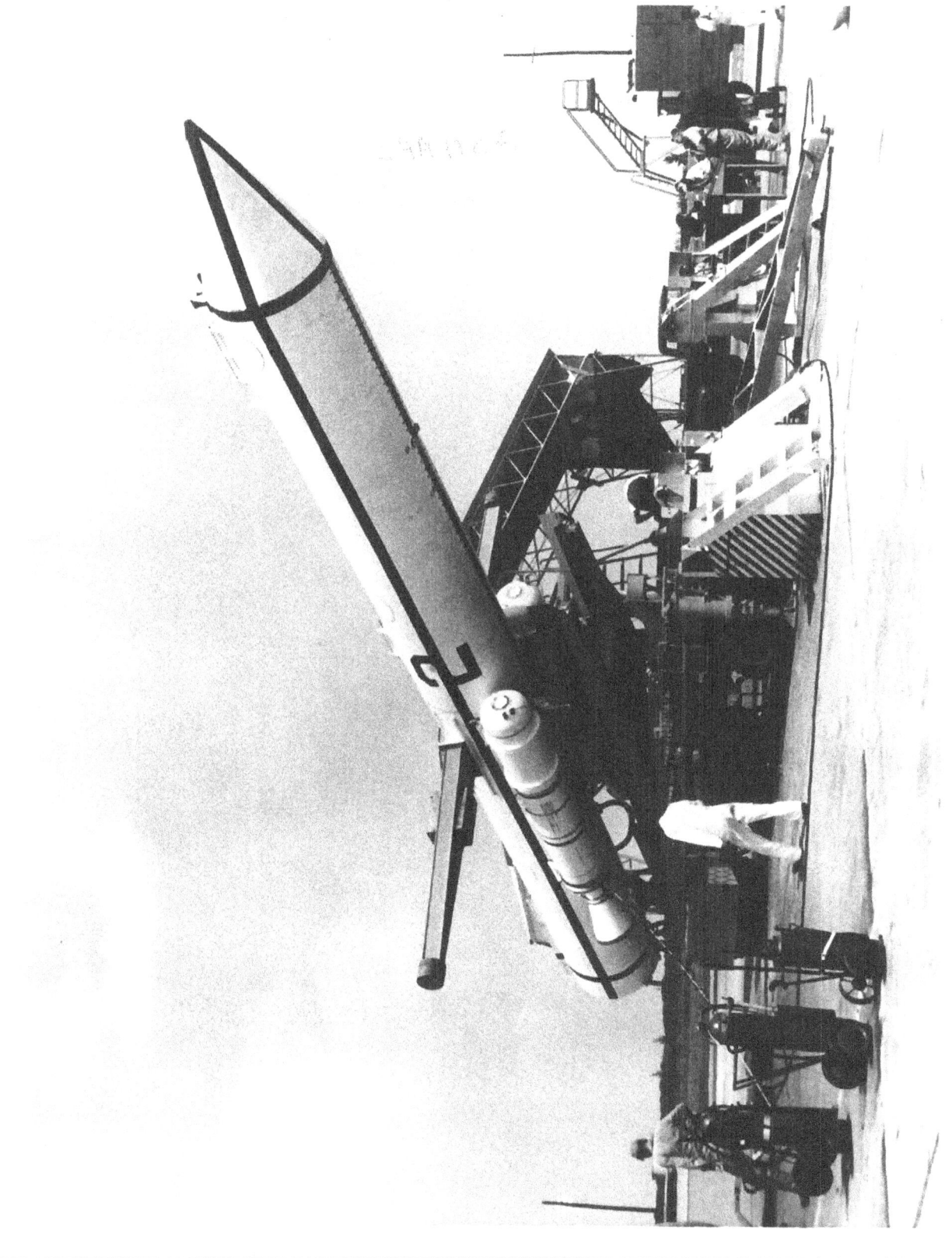

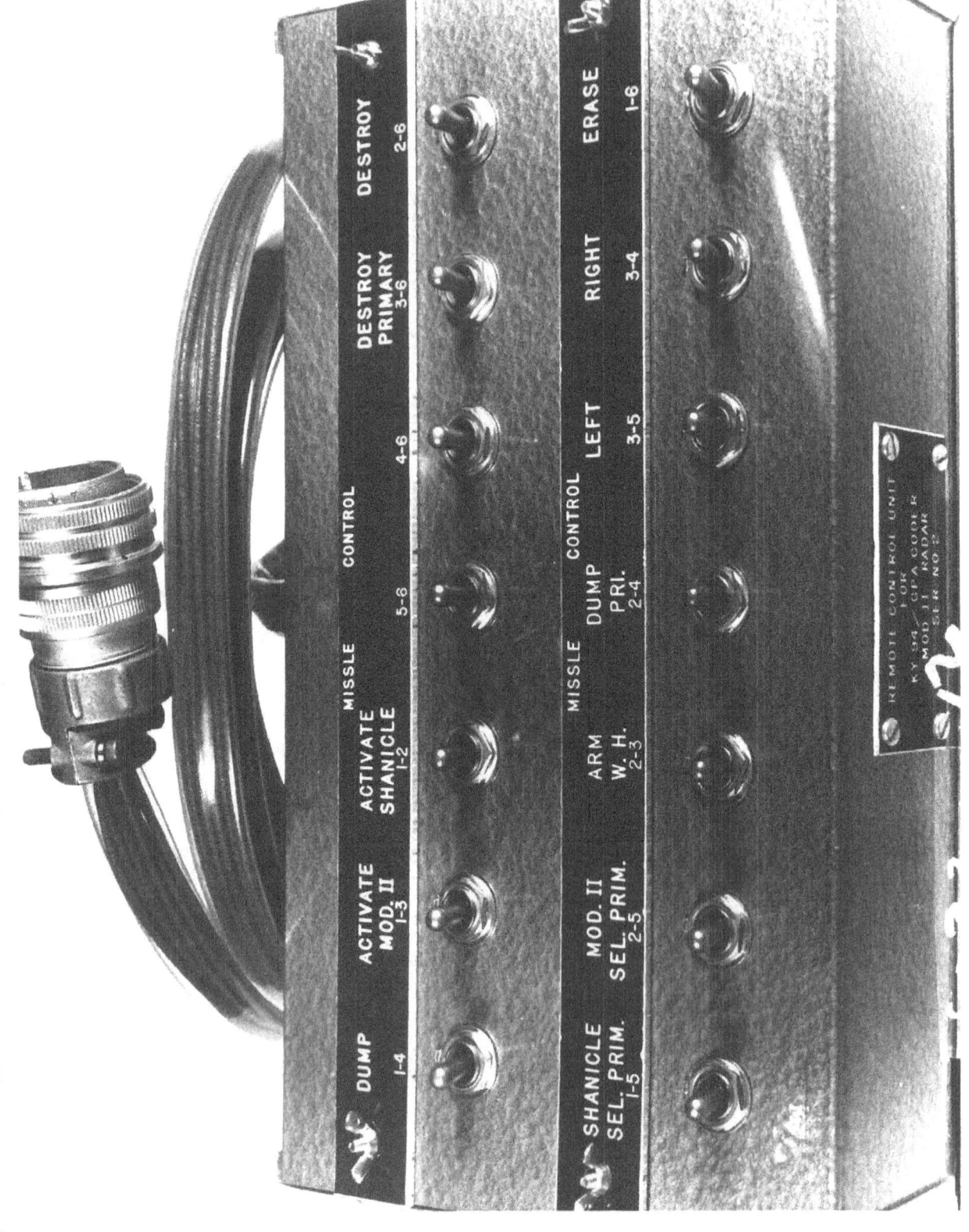

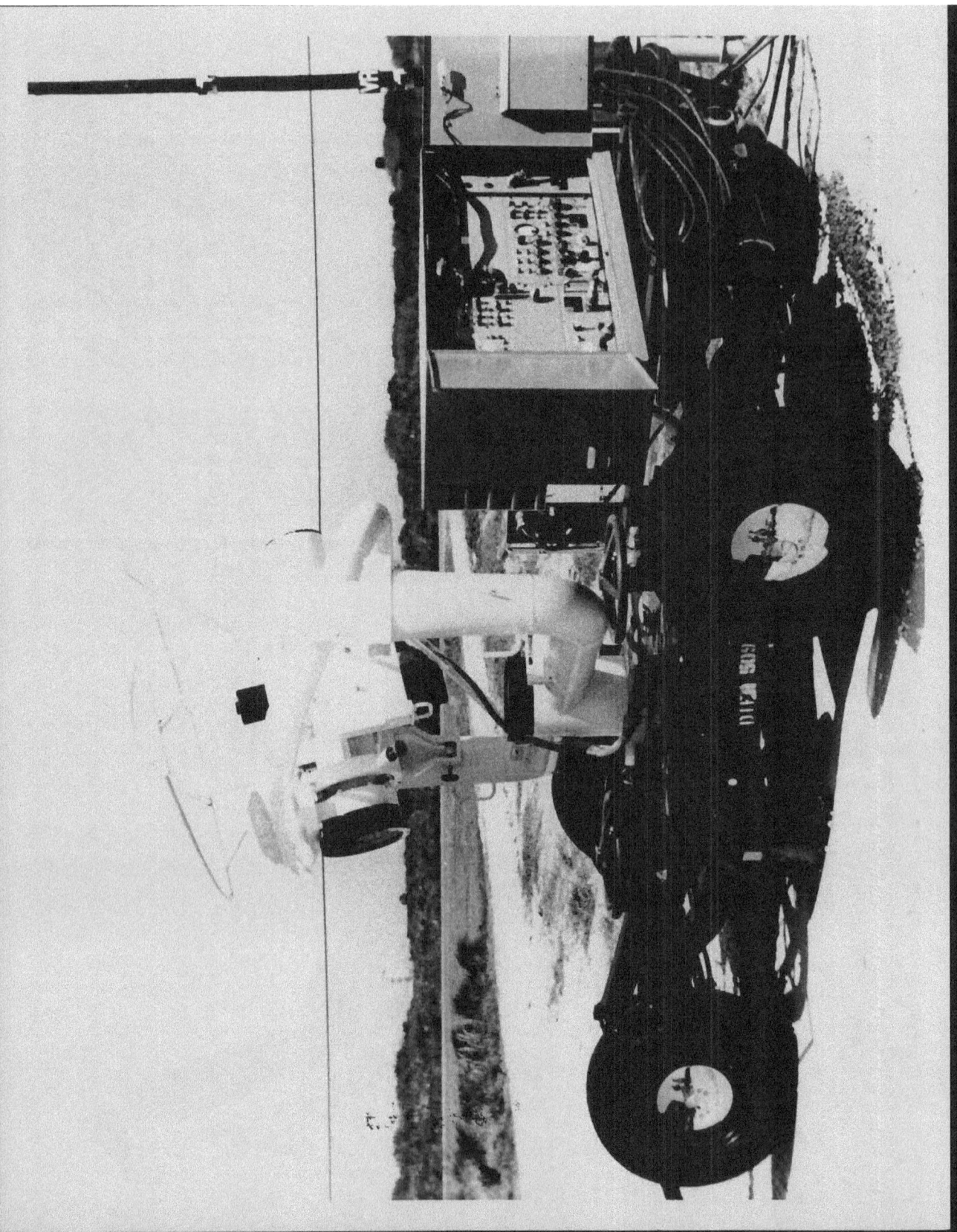

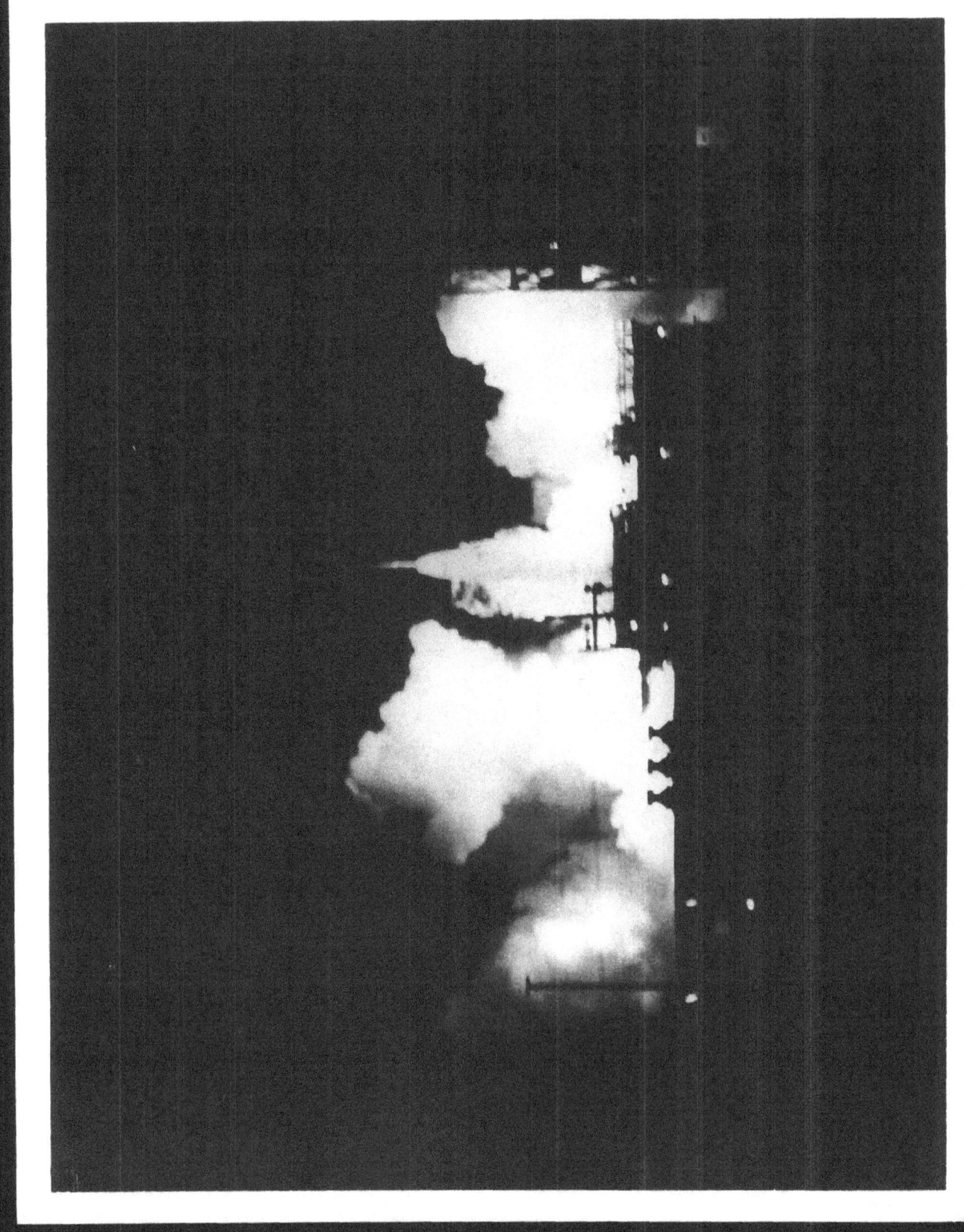

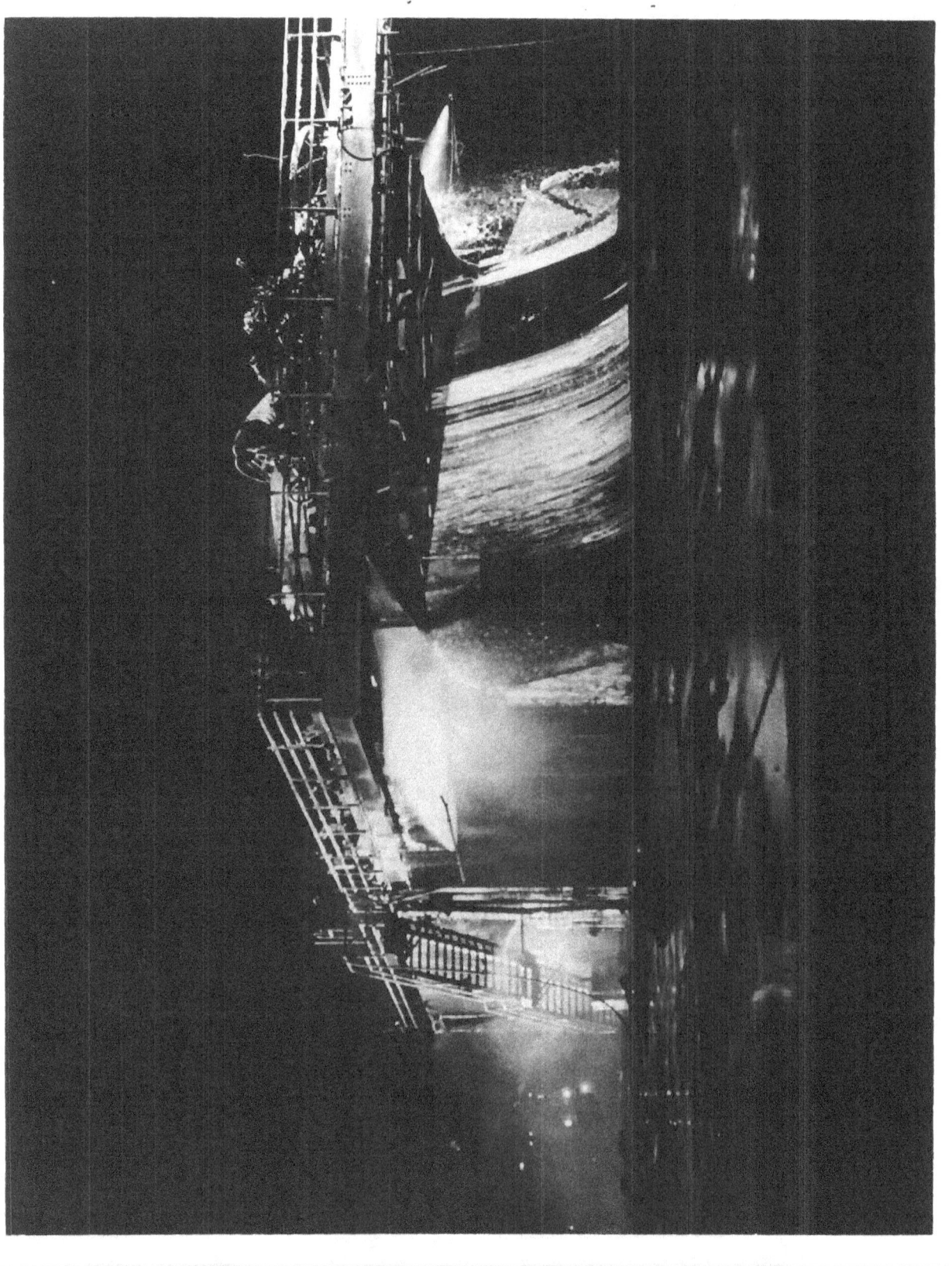

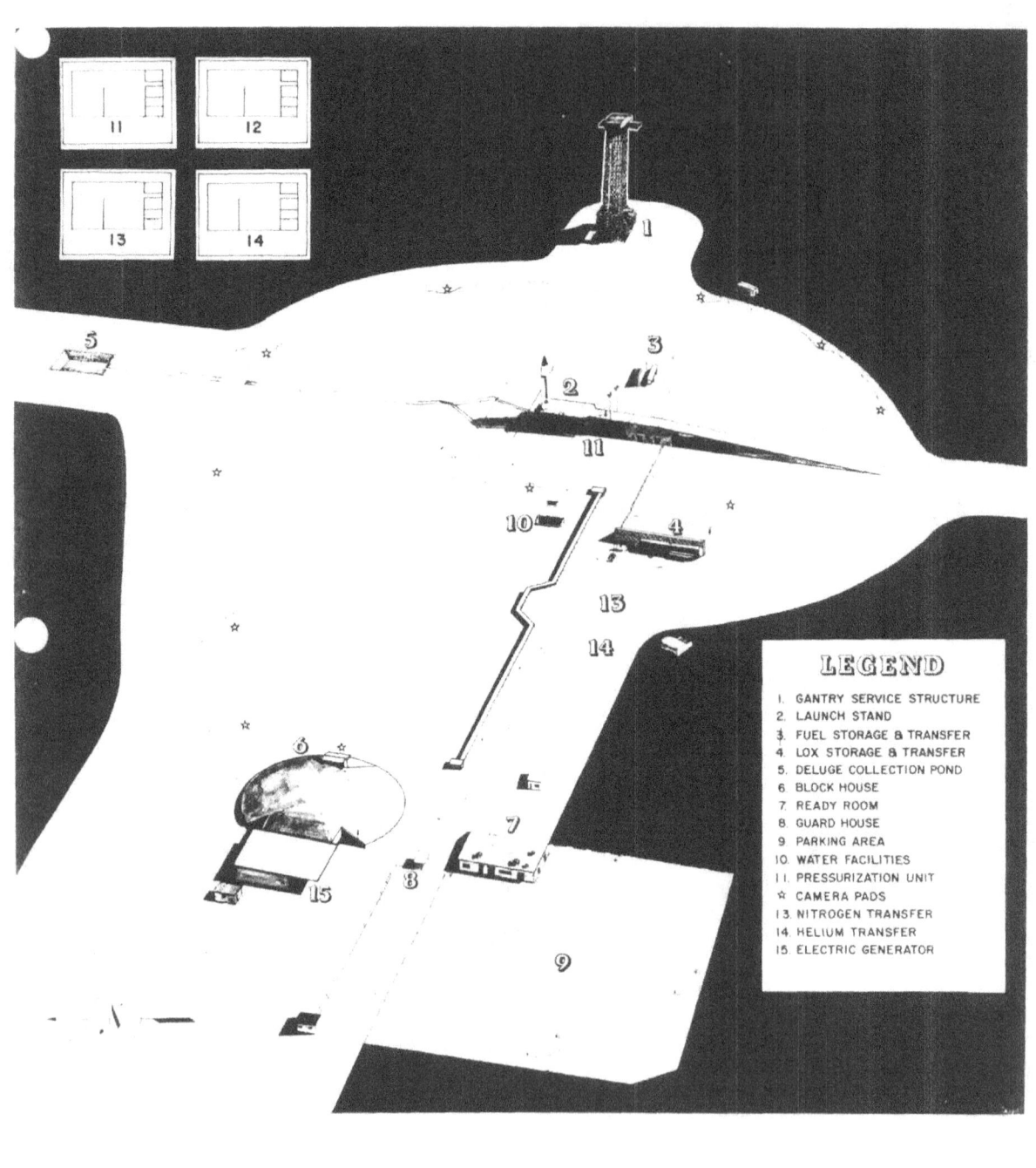

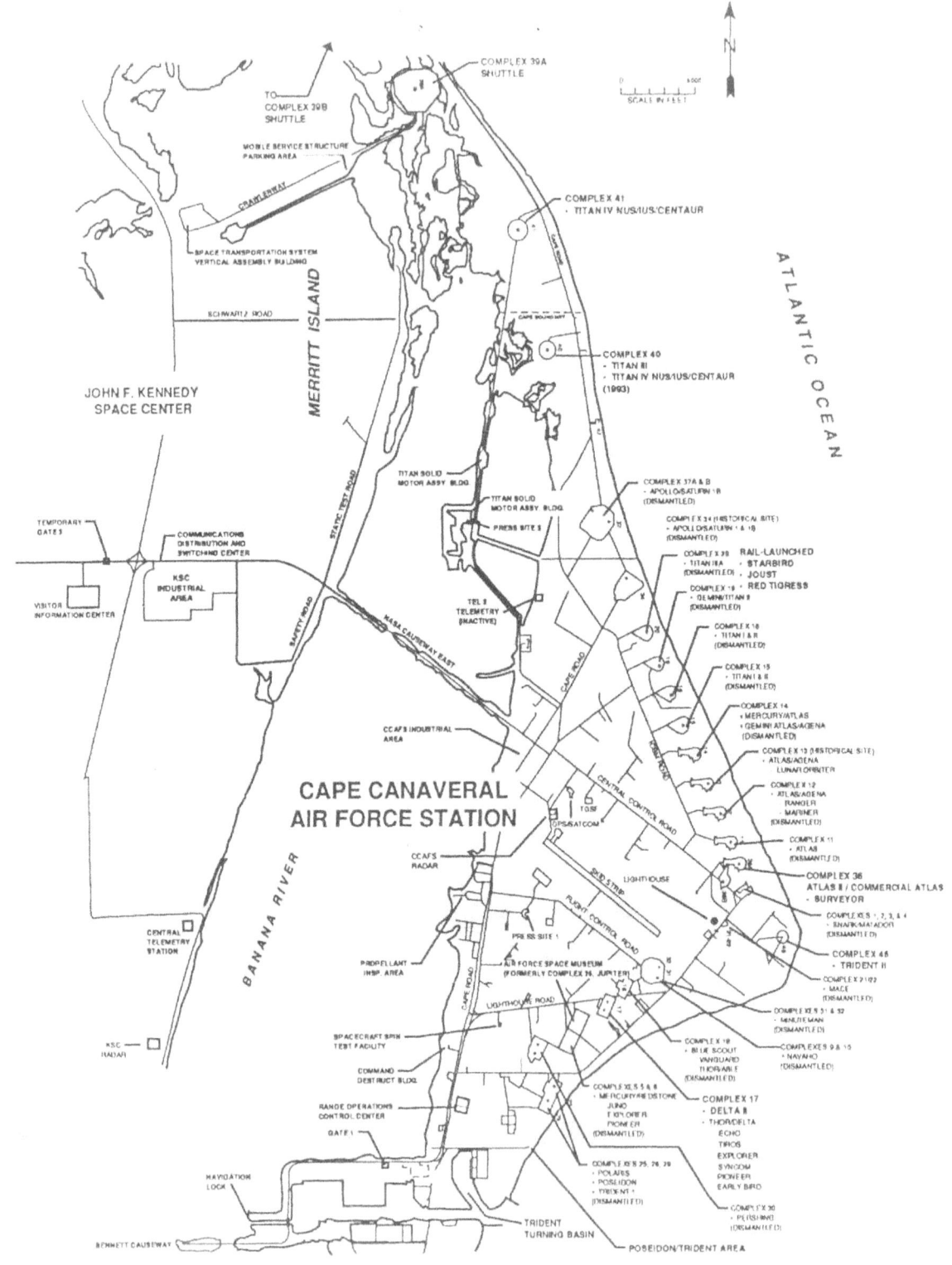

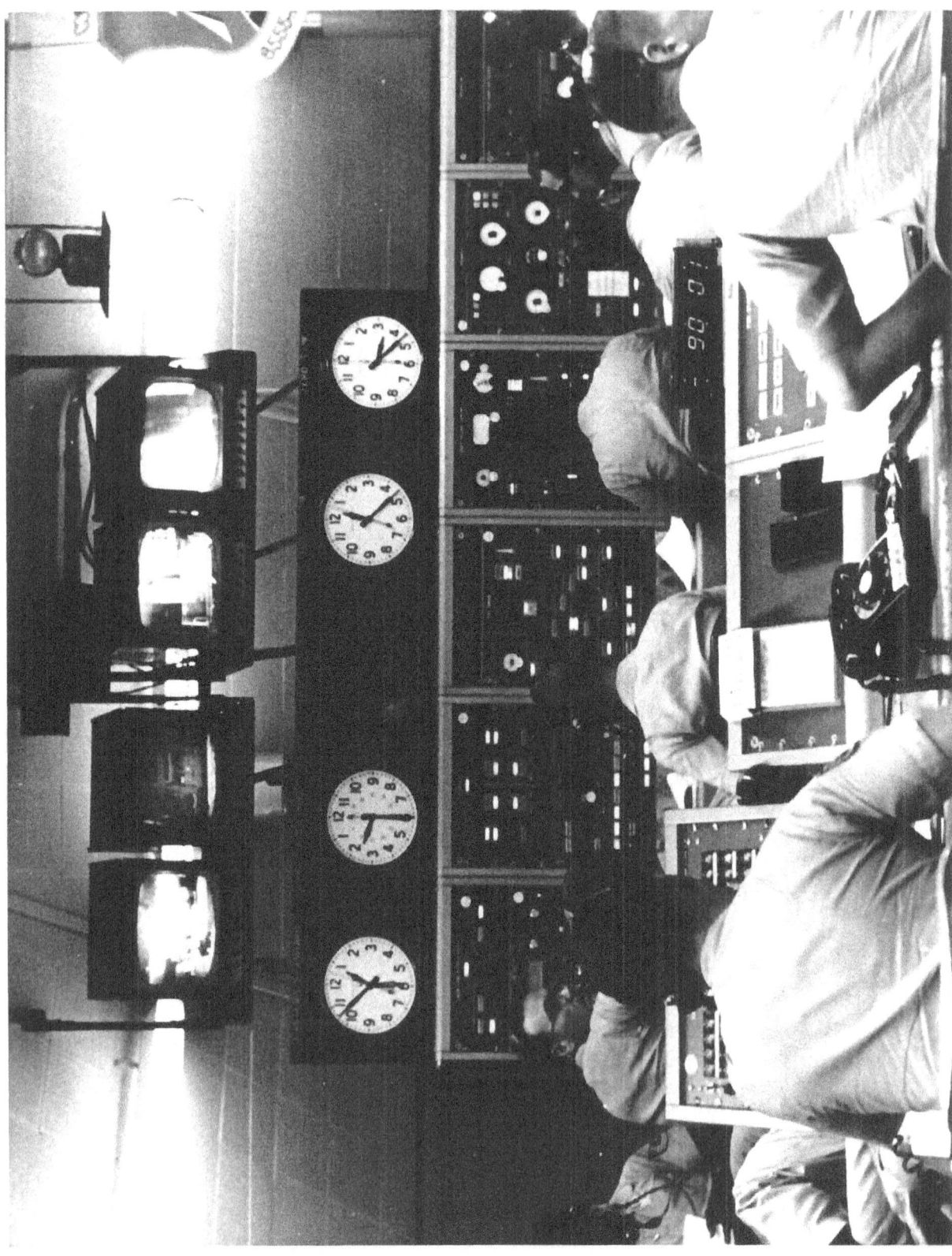

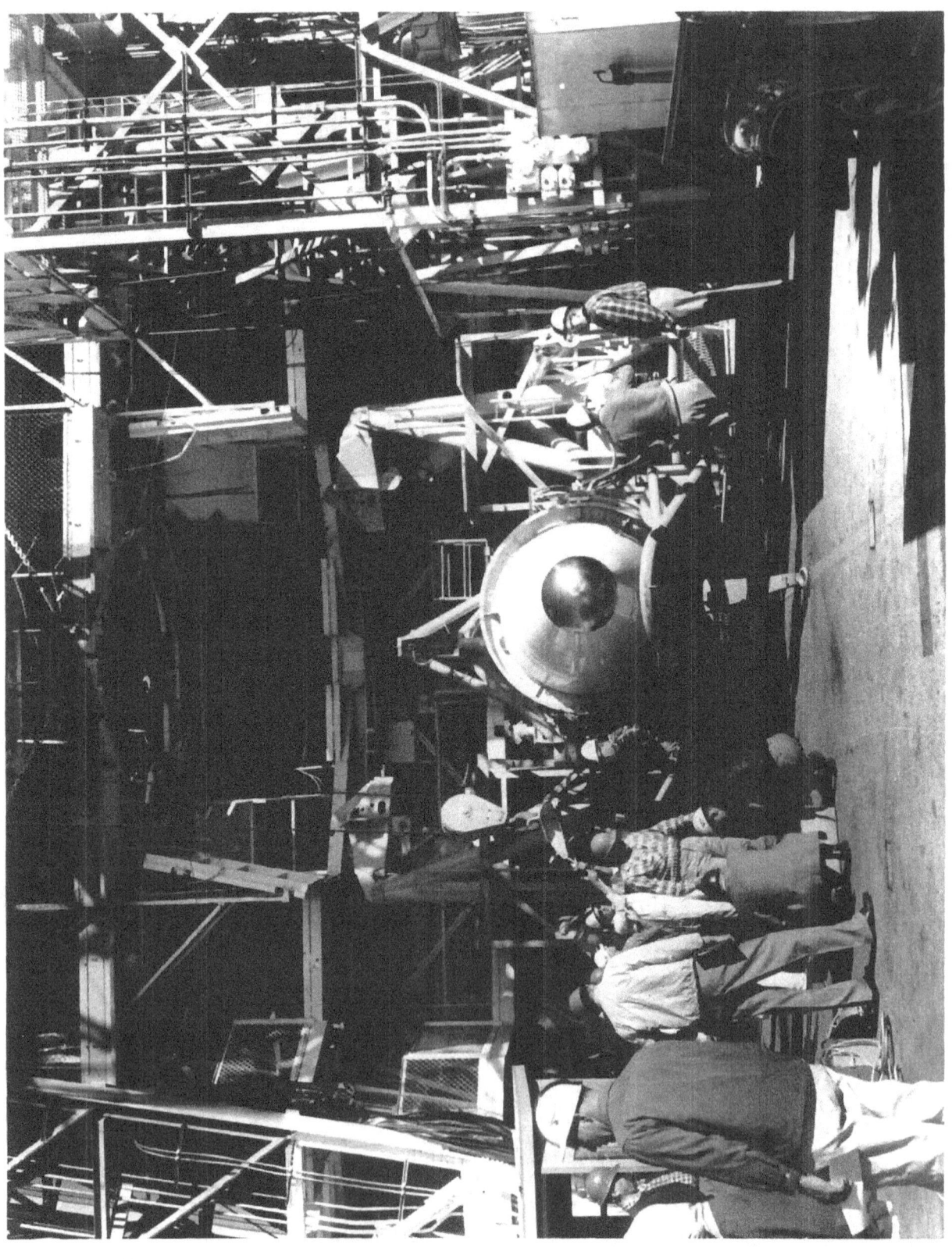

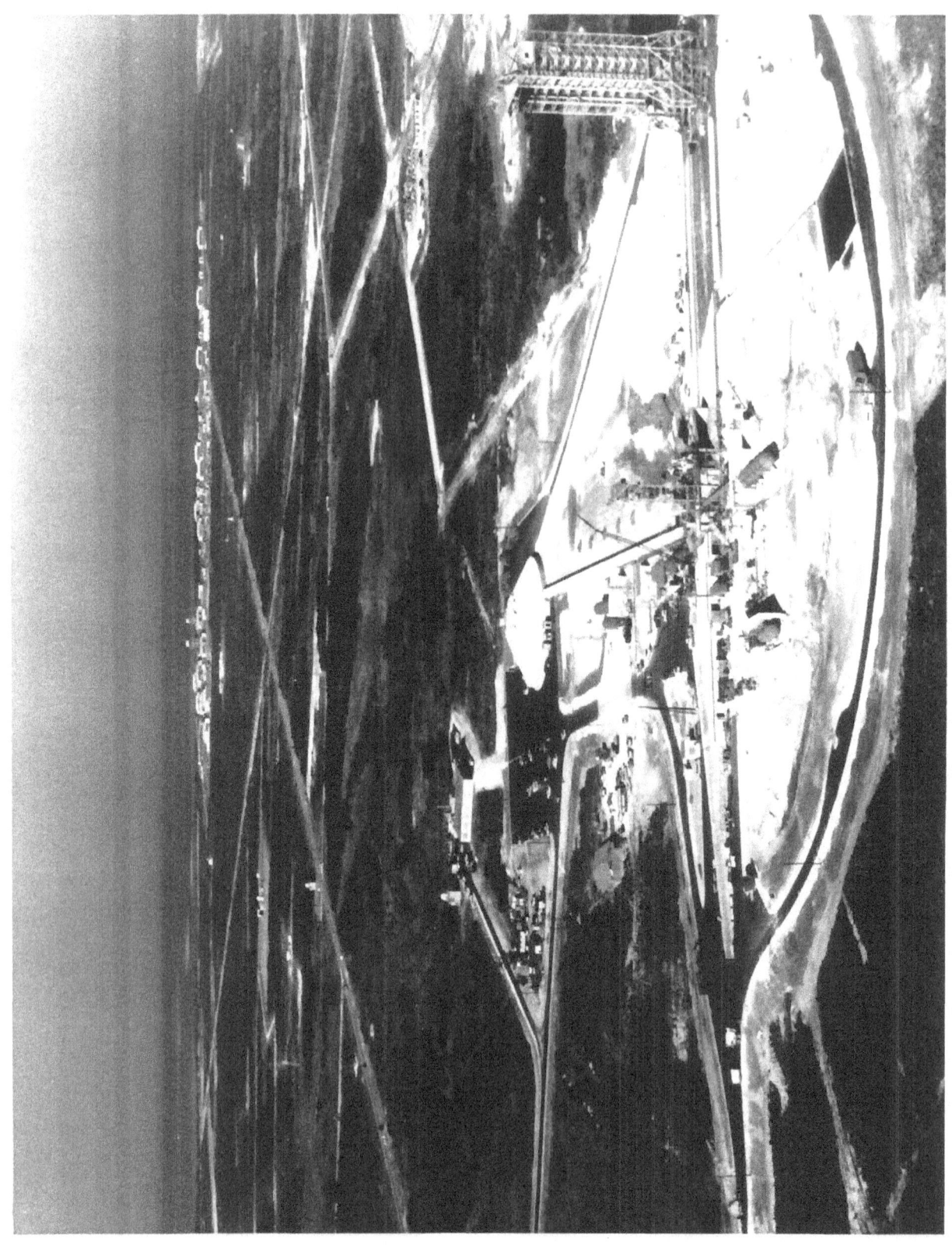

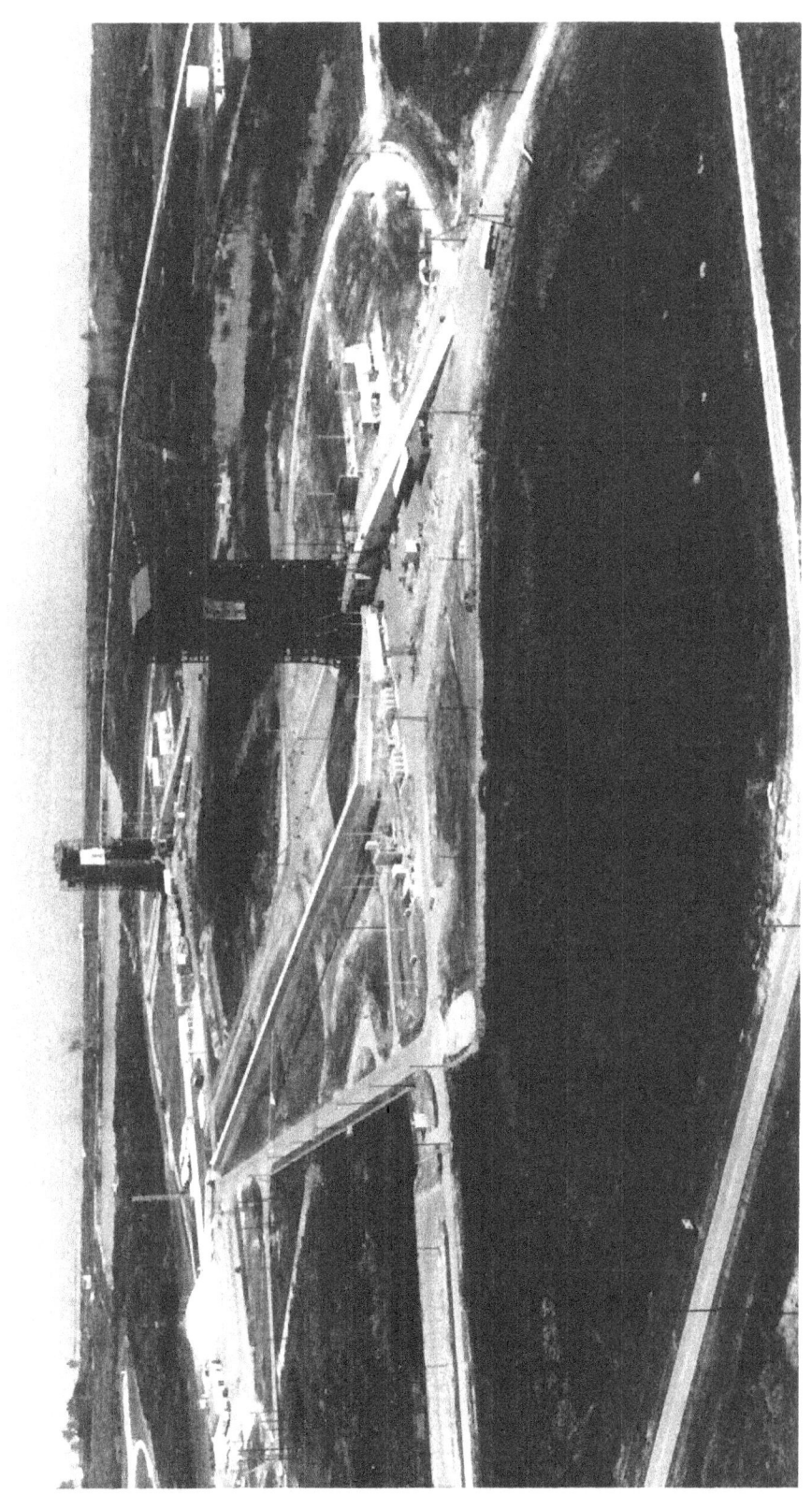

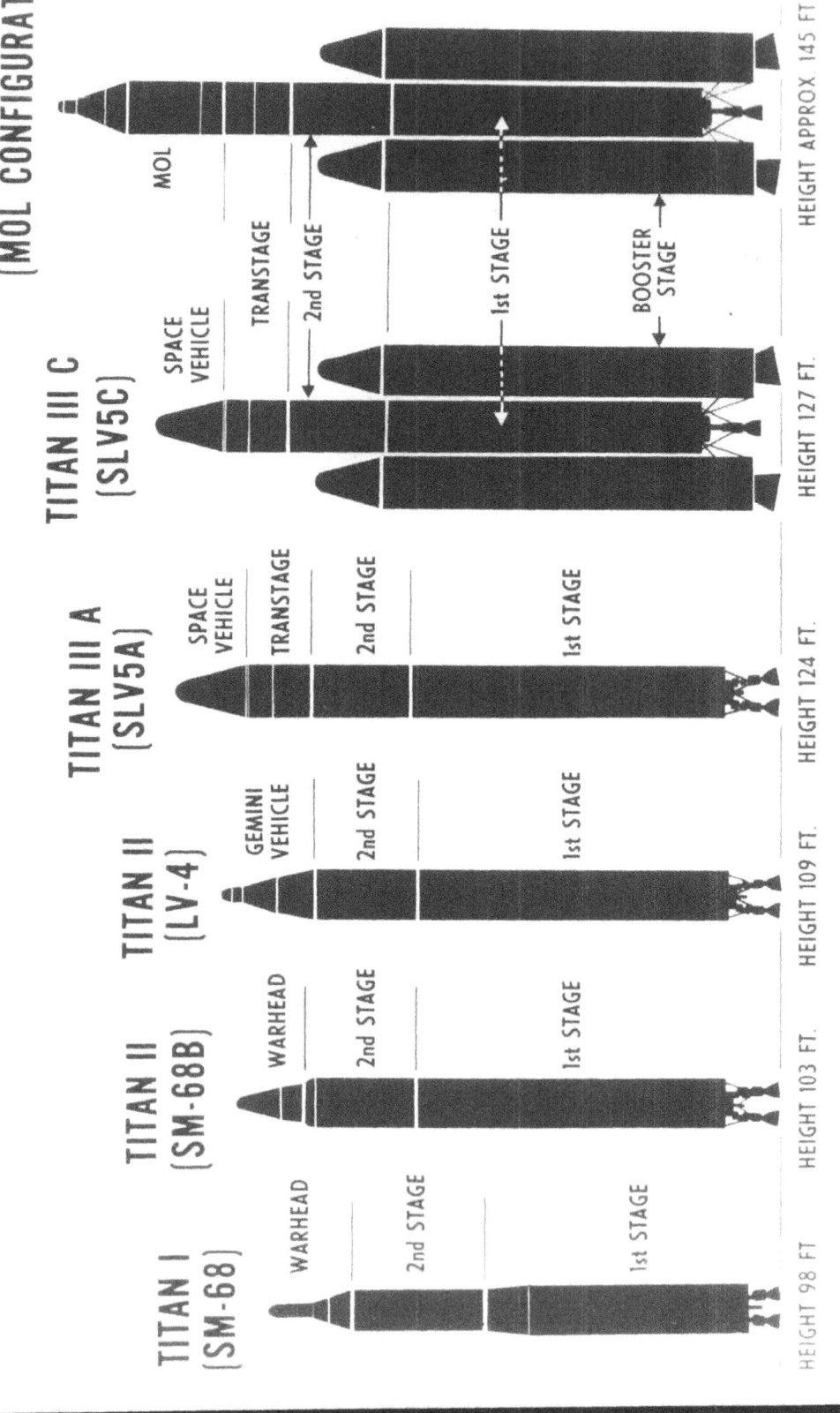

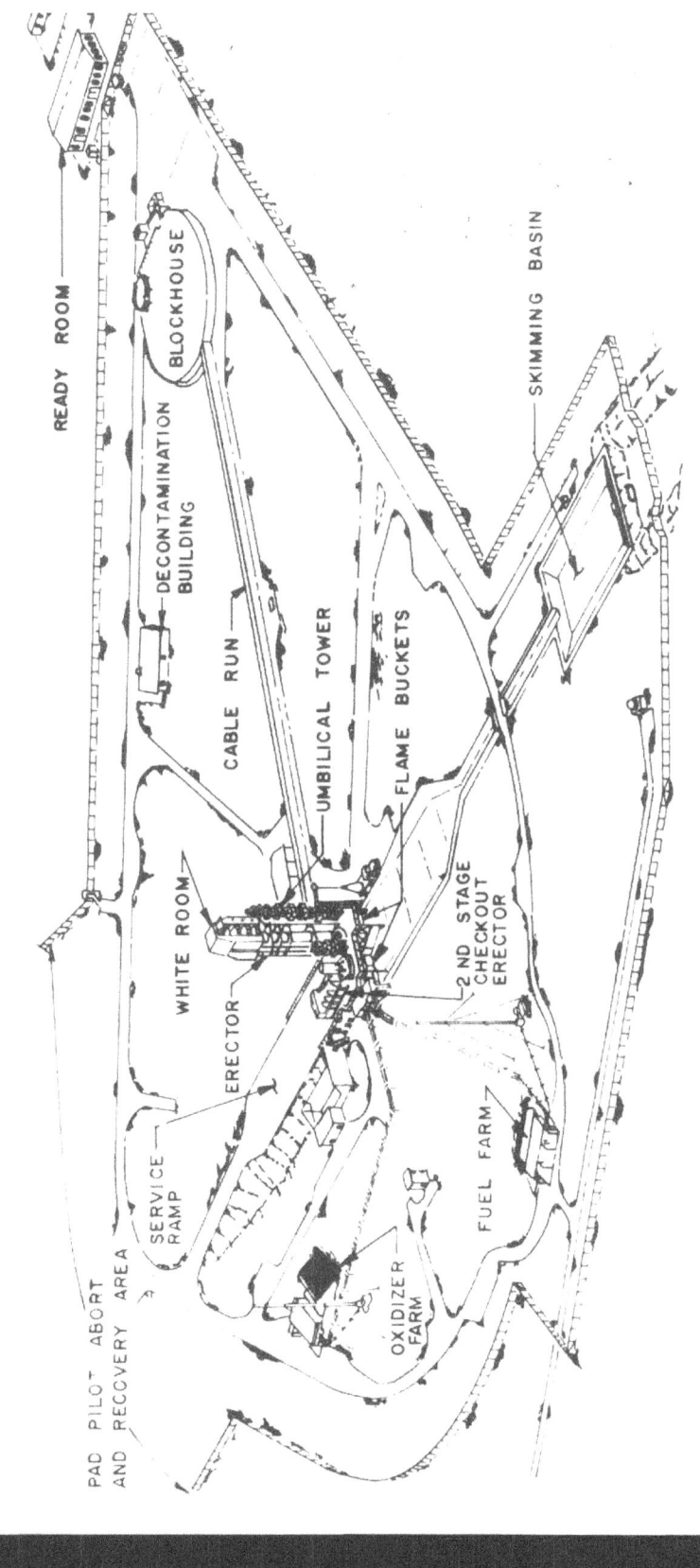

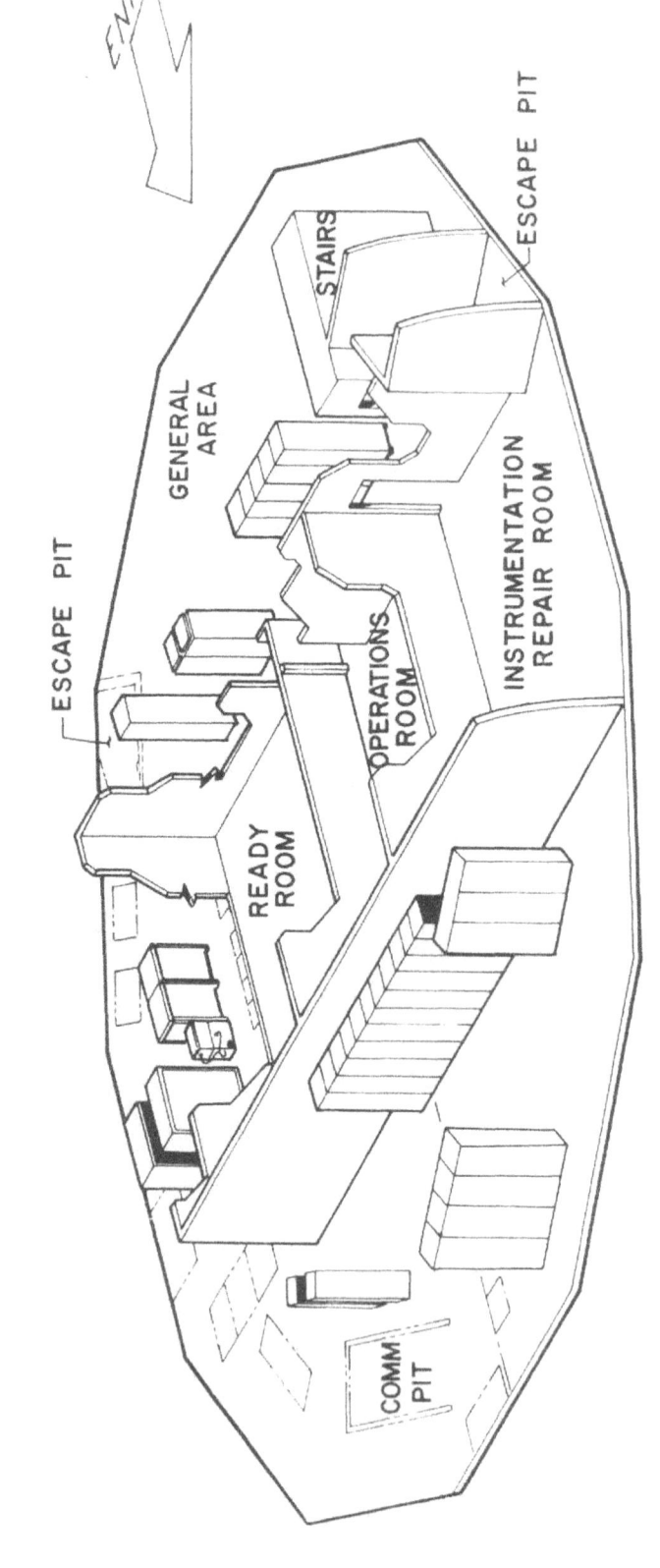

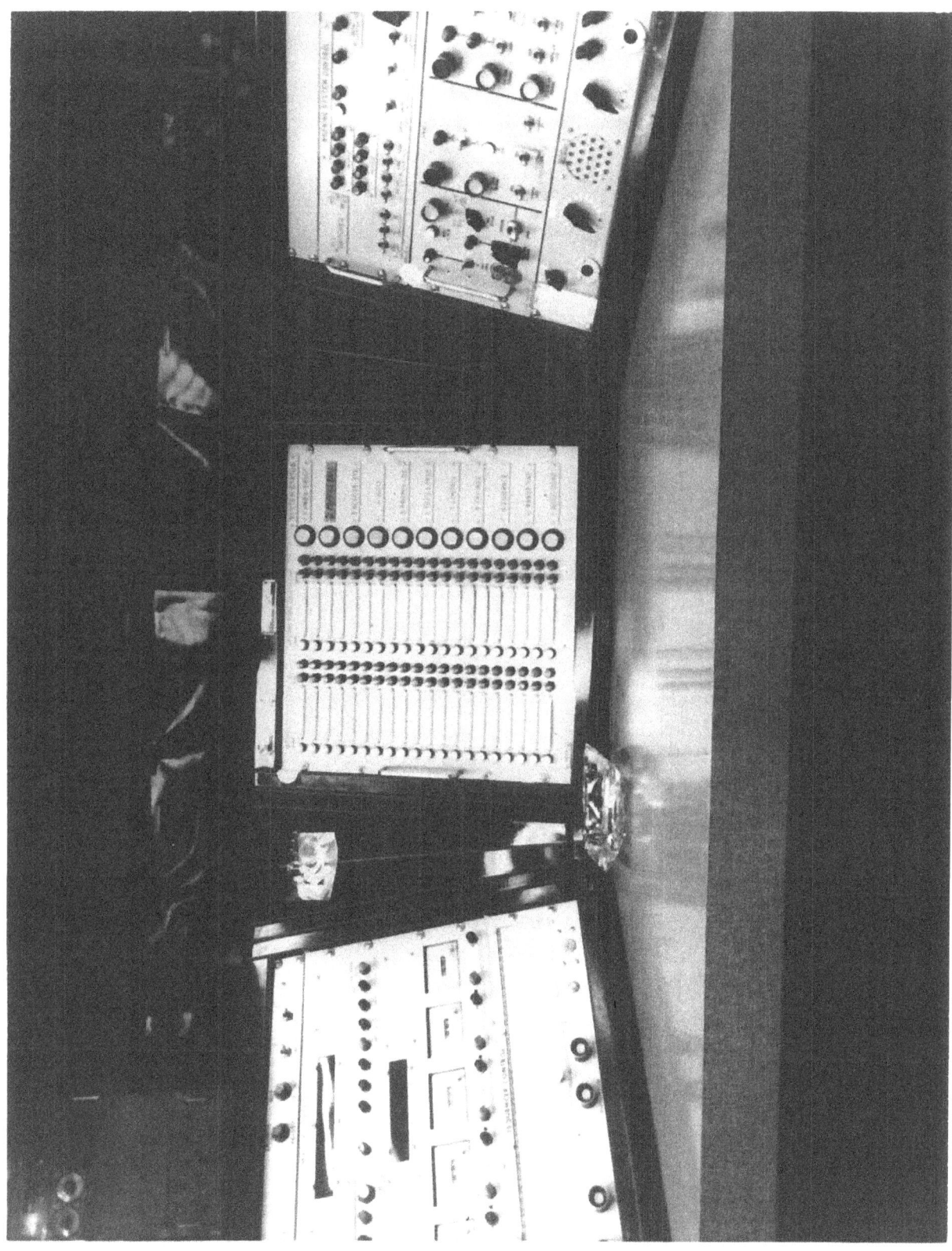

COMPLEX 19
GEMINI TITAN II
THE UNITED STATES 2 MAN SPACE MISSIONS

Mission	Orbits	Date	Crew
GT-3	4 ORBITS	23 MAR. 1965	MAJ. VIRGIL I. GRISSOM, USAF / LT. CDR. JOHN W. YOUNG, USN
GT-4	62 ORBITS	3-7 JUNE 1965	MAJ. JAMES A. McDIVITT, USAF / MAJ. EDWARD H. WHITE II, USAF
GT-5	120 ORBITS	21-29 AUG. 1965	LT. COL. L. GORDON COOPER, JR., USAF / LT. CDR. CHARLES CONRAD JR., USN
GT-7	220 ORBITS	4-18 DEC. 1965	COL. FRANK BORMAN, USAF / CAPT. JAMES A. LOVELL, USN
GT-6	16 ORBITS	15-16 DEC. 1965	CAPT. WALTER M. SCHIRRA, USN / LT. COL. THOMAS P. STAFFORD, USAF
GT-8	7 ORBITS	16 MAR. 1966	MR. NEIL A. ARMSTRONG / MAJ. DAVID R. SCOTT, USAF
GT-9	48 ORBITS	3-6 JUNE 1966	LT. COL. THOMAS P. STAFFORD, USAF / LT. CDR. EUGENE A. CERNAN, USN
GT-10	47 ORBITS	18-21 JULY 1966	CDR. JOHN W. YOUNG, USN / MAJ. MICHAEL COLLINS, USAF
GT-11	47 ORBITS	12-15 SEPT. 1966	CDR. CHARLES CONRAD JR., USN / LT. CDR. RICHARD F. GORDON JR., USN

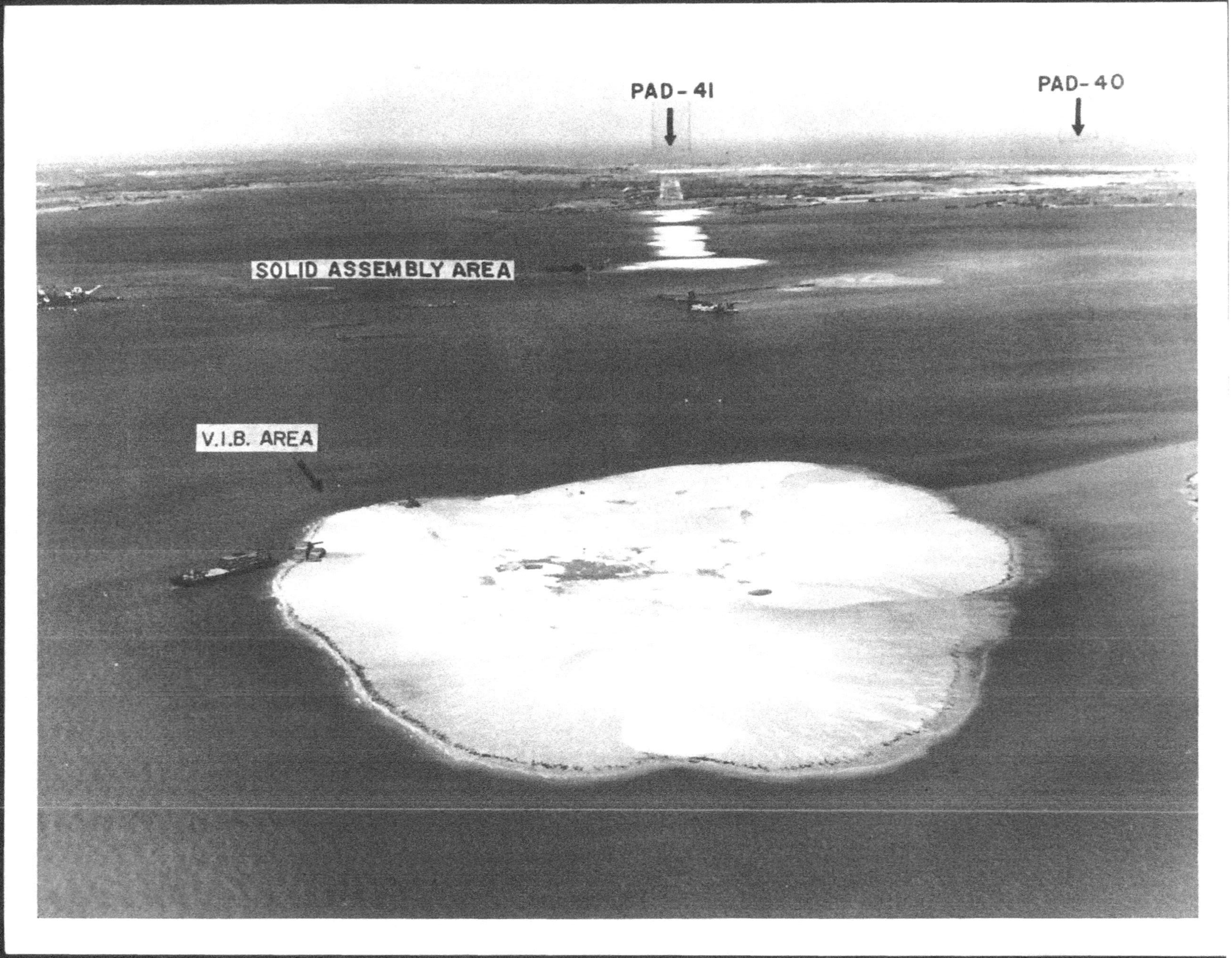

PAD-40

PAD-41

SOLID ASSEMBLY AREA

V.I.B. AREA

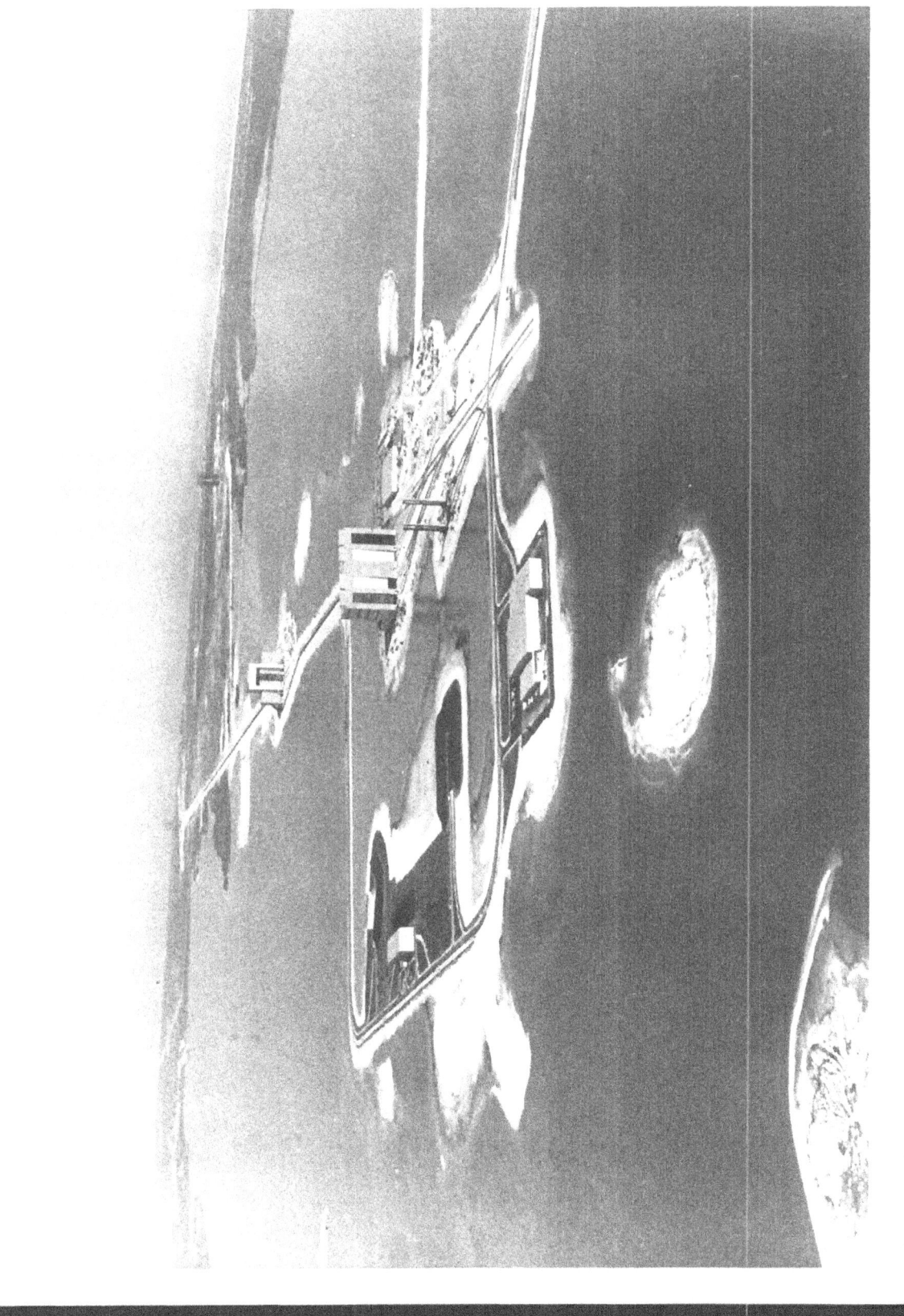

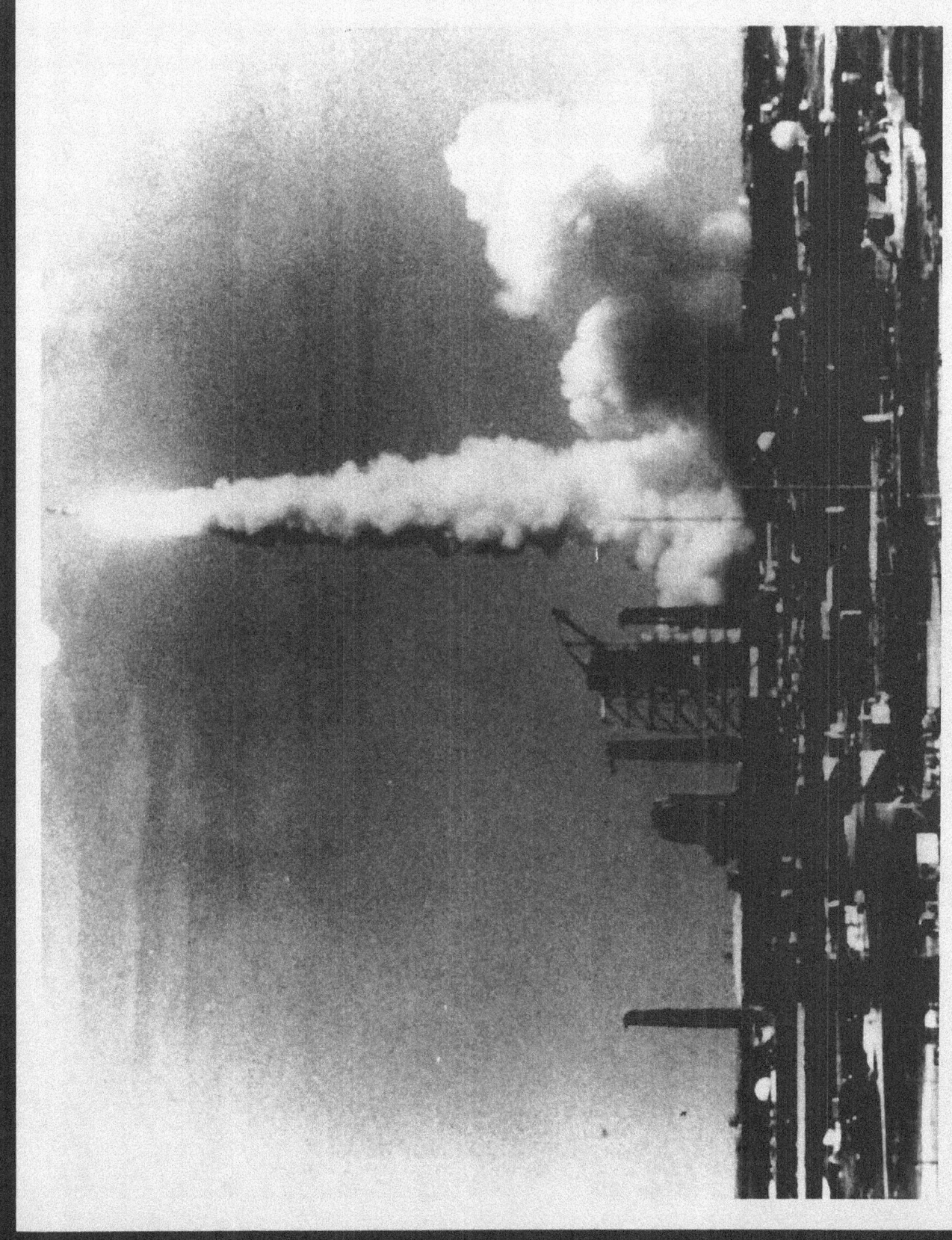

www.ingramcontent.com/pod-product-compliance
Lightning Source LLC
Chambersburg PA
CBHW082109230426
43671CB00015B/2643